Home Security Systems DIY using Android and Arduino

Robert Chin

Copyright © 2015 Robert Chin

All rights reserved.

Table of Contents

About the Author 4

Introduction 5

Chapter 1: Introducing the Arduino 6

Chapter 2: Arduino Programming Language Basics 26

Chapter 3: The Android Controller and Bluetooth Communication with Arduino 42

Chapter 4: Simple Wireless Intruder Alarm System with Motion Detector 89

Chapter 5: Hands on Example: Creating a Simple Intruder Alarm System 147

Chapter 6: ArduCAM Mini Wireless Intruder Alarm/Video Surveillance System 160

Chapter 7: Hands on Example: Building an ArduCAM Intruder Alarm / Surveillance System ... 226

Chapter 8: Deploying your Wireless Intruder Alarm and Surveillance System 252

About the Author:

Robert Chin has a Bachelor of Science degree in computer engineering and is experienced in Android development, Arduino development, C/C++, Unreal Script, Java, DirectX, OpenGL, and OpenGL ES 2.0. He has written 3d games for the Windows, and Android platforms. He is the author of "Beginning Arduino ov7670 Camera Development". He is also the author of "Beginning Android 3d Game Development", and "Beginning IOS 3d Unreal Games Development" both published by Apress and was the technical reviewer for "UDK Game Development" published by Course Technology CENGAGE Learning.

Introduction

This book shows you how to construct your own home security systems using an Android device, Arduino microcontroller, a infrared motion detector, an Arduino compatible Bluetooth adapter, and an optional ArduCAM digital camera. It also gives in depth details so that you can modify and customize the security systems yourself.

> Important Note: Chapter 5 and Chapter 7 are the "Hands On Example" quick start guides that show you how to quickly build these home security systems for use. Chapter 8 covers some useful tips for the deployment and construction of your alarm systems.

A summary of the content of the book's chapters follows.

Chapter 1: Introducing the Arduino - In this chapter I introduce you to the Arduino. I first give a brief explanation of what the Arduino is. I then specifically concentrate on the Arduino Uno.

Chapter 2: Arduino Programming Language Basics - In this chapter I go over the basics of the Arduino Language.

Chapter 3: The Android Controller and Bluetooth Communication with Arduino - In this chapter I give an overview of Android development and Bluetooth communication including detailed explanations on the Arduino and Android code and hardware required to implement Bluetooth.

Chapter 4: Simple Wireless Intruder Alarm System with Motion Detector - This chapter covers a simple wireless intruder alarm system using the Android, Arduino, a motion detector, and a bluetooth adapter. The Android and Arduino code which implements this alarm system is discussed in detail.

Chapter 5: Hands on Example: Creating a Simple Intruder Alarm System - In this chapter I give a quick start guide to setting up the simple intruder alarm system I discussed in detail in Chapter 4. I also present a quick start guide on how to actually use this system.

Chapter 6: ArduCAM Mini Wireless Intruder Alarm/Video Surveillance System - This chapter covers the ArduCAM based intruder alarm. The Android and Arduino code which implements this alarm system is discussed in detail.

Chapter 7: Hands on Example: Building an ArduCAM Intruder Alarm / Surveillance System - In this chapter I show you how to build an intruder/surveillance alarm system that uses the ArduCAM digital camera described in Chapter 6. I also give a quick start guide presentation on how to use this type of alarm system.

Chapter 8: Deploying your Wireless Intruder Alarm and Surveillance System - This chapter will discuss deployment of your new intruder alarm/surveillance system.

Chapter 1

Introducing the Arduino

In this chapter I introduce you to the Arduino. I first give a brief explanation of what the Arduino is. I then specifically concentrate on the Arduino Uno. I discuss the general features of the Arduino Uno including the capabilities and key functional components of the device. Next, I discuss the Arduino IDE (Integrated Development Environment) software that is needed to develop programs for the Arduino. I cover each key function of the Arduino IDE and then conclude with a hands on example where I give detailed step by step instructions on how to set up the Arduino for development and how to run and modify an example program using the Arduino IDE.

What is an Arduino?

The Arduino is an open source microcontroller that uses the C and C++ languages to control digital and analog outputs to devices and electronics components and to read in digital and analog inputs from other devices and electronics components for processing. For example, the Arduino can read in information from a sensor to a home security system that would detect the heat that a human being emits and sends a signal to the Arduino to indicate that a human is in front of the sensor. After receiving this information the Arduino can send commands to a camera such as the ArduCAM Mini digital camera to start taking pictures of the intruder or intruders. There are many different Arduino models out there. However, in order to perform the examples in this book you will need an Arduino model with enough pins to connect the components you desire such as a camera, blue tooth adapter, and/or motion sensor. The official Arduino logo is shown in Figure 1-1.

Figure 1-1. Official Arduino Logo

Note: The official web site of the Arduino project is http://www.arduino.cc however in 2014 there appears to be a split between the founders of the Arduino project as to who controls the "Arduino" trademark name. Another web site called http://www.Arduino.org

was created by the company of one of the Arduino founders that split from the main group.

The Arduino UNO

There is a wide range of Arduino models ranging from models that are small and can actually be worn by the user to Arduino models with many digital and analog input/output pins. For the examples in this book I recommend the Arduino Uno. The Arduino Uno is an open source microcontroller that has enough digital ports to accommodate the camera and a sd card reader/writer with enough digital and analog ports for other devices, sensors, lights, and any other gadgets that you may require for your own custom projects. There is an official Arduino Uno board made by a company called Arduino SRL formerly Smart Projects formed by one of the founders of the Arduino. See Figure 1-2.

Figure 1-2. The Official Arduino UNO

There are also unofficial Arduino Uno boards made by other companies. A good way to tell which board is official and which is unofficial is by the color of a component that is located near the Arduino's usb port. The component on official Arduino boards is colored a metallic gold. The component on unofficial boards has a green color. The writing on the component also differs. See Figure 1-3.

8

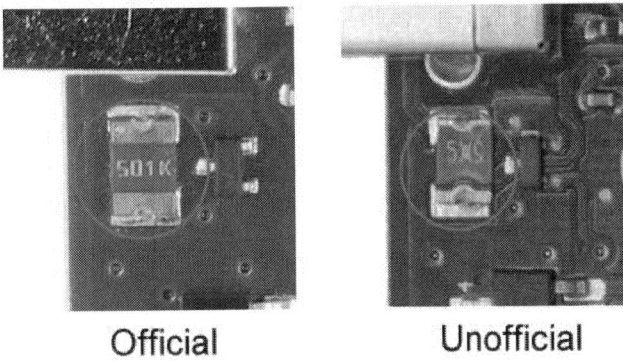

Figure 1-3 Official vs. Unofficial Arduino Boards

There are also other companies that manufacture Arduino Uno boards. See Figure 1-4. Also note the green component next to the usb port. Since the Arduino schematics are open source and other companies can legally manufacture this board there are many competing companies making this board and the boards vary in quality and price. Generally, an unofficial Arduino Uno is around $10 and an official Arduino Uno board is around $20.

Figure 1-4. Unofficial Arduino UNO

The Arduino UNO Specifications

Microcontroller   ATmega328P

Operating Voltage   5V

Input Voltage (recommended)	7-12V
Input Voltage (limit)	6-20V
Digital I/O Pins	14 (of which 6 provide PWM output)
PWM Digital I/O Pins	6
Analog Input Pins	6
DC Current per I/O Pin	20 mA
DC Current for 3.3V Pin	50 mA
Flash Memory	32 KB (ATmega328P)
of which 0.5 KB used by bootloader	
SRAM	2 KB (ATmega328P)
EEPROM	1 KB (ATmega328P)
Clock Speed	16 MHz
Length	68.6 mm
Width	53.4 mm
Weight	25 g

Arduino UNO Components

This section covers the functional components of the Arduino UNO.

USB Connection Port

The Arduino Uno has a USB connector that is used to connect the Arduino to the main computer development system via standard USB A male to B male cable so it can be programmed and debugged. See Figure 1-5.

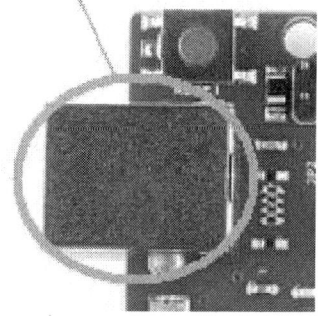

Figure 1-5. USB Connector

9V Battery Connector

The Arduino Uno has a 9-volt battery connector where you can attach a 9-volt battery to power the Arduino. See Figure 1-6.

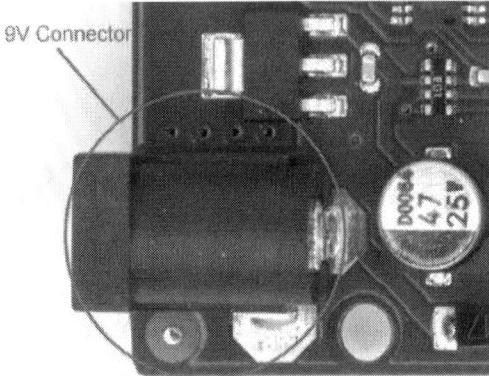

Figure 1-6. 9 volt battery connector

Reset Button

There is a reset button on the Arduino Uno where you can press the button down to reset the board. This restarts the program contained in the Arduino's memory. See Figure 1-7.

Figure 1-7. Reset Button

Digital Pins

The Arduino Uno has many digital pins capable of simulating analog output through the process of pulse width modulation. For example, an L.E.D. light generally has only two modes which is on (full brightness) or off (no light emitted). However, with digital pulse width modulation the L.E.D. light can appear to have a brightness in between on and off. For instance, with PWM (Pulse Width Modulation) an L.E.D. can start from an off state and slowly brighten until it is at its highest brightness level and then slowly dim until back to the off state. The digital pins on the Arduino Uno are pins 0 through pin 13. These PWM capable digital pins are circled in Figure 1-8.

Figure 1-8. Digital pins

Communication

The communication section of the Arduino Uno contains pins for serial communication between the Arduino and another device such as a Bluetooth adapter or your personal computer. The Tx0 and the Rx0 pins are connected to the USB port and serve as communication from your Arduino to your computer through your USB cable. The Serial Monitor that can be used for sending data to the Arduino and reading data from the Arduino uses the Tx0 and Rx0 pins. Thus, you should not connect anything to these pins if you want to use the Serial Monitor to debug your Arduino programs or to receive user input. I will talk more about the Serial Monitor later in this book. See Figure 1-9.

Figure 1-9. Serial Communication

The I2C interface consists of an SDA pin which is pin 4 and is used for data and an SCL pin which is pin 5 and is used for clocking or driving the device or devices attached to the I2C interface. The SDA and SCL pins are circled in Figure 1-10.

Figure 1-10. I2C Interface

Analog Input

The Arduino Uno has 6 analog input pins that can read in a range of values instead of just digital values of 0 or 1. The analog input pin uses a 10 bit analog to digital converter to transform voltage input in the range of 0 volts to 5 volts into a number in the range between 0 to 1023. See Figure 1-11.

Figure 1-11. Analog Input

Power

The Arduino Uno has outputs for 3.3 volts and 5 volts. One section that provides power is located on the side of the Arduino. You can also provide your own power source by connecting the positive terminal of the power source to the Vin pin and the ground of the power source to the Arduino's ground. Make sure the voltage being supplied is within the Arduino board's voltage range. See Figure 1-12.

Figure 1-12. 3.3 volt and 5 volt Power outputs

The ground connections on the Arduino Uno are shown circled in Figure 1-13.

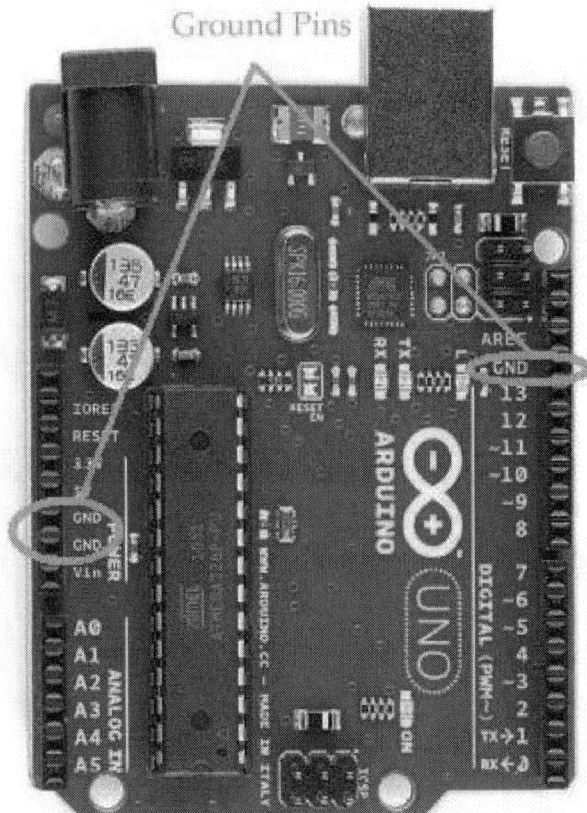

Figure 1-13. Arduino Uno Ground Connections

Arduino Development System Requirements

Developing projects for the Arduino can be done on the Windows, Mac, and Linux operating systems. The software needed to develop programs that run on the Arduino can be downloaded from the main web site at:

http://www.arduino.cc/en/Main/Software

The following is a summary of the different types of Arduino IDE distributions that are available for download. You will only need to download and install one of these files. The file you choose will depend on the operating system your computer is using.

Windows

- Windows Installer – This is a .exe file that must be run to install the Arduino Integrated Development Environment.

- Windows ZIP file for non admin install – This is a zip file that must be uncompressed in order to install the Arduino Integrated Development Environment. 7-zip is a free file compression and uncompression program available at http://www.7-zip.org that can be used to uncompress this program.

Mac

- Mac OS X 10.7 Lion or newer – This is a zip file that must be uncompressed and installed for users of the Mac operating system

Linux

- Linux 32 bits – Installation file for the Linux 32 bit operating system.

- Linux 64 bits – Installation file for the Linux 64 bit operating system.

The easiest and cheapest way to start Arduino development is probably through using the Windows version on an older operating system such as Windows XP. In fact, the examples in this book were created by using the Windows version of the Arduino IDE running on Windows XP. There are in fact many sellers on Ebay where you can purchase a used Windows XP computer for around $50-$100. So if you are starting from scratch and are looking for a inexpensive development system for the Arduino then consider buying a used Windows XP based computer. The only caution is that support for the Windows XP has ended in the United States and some other parts of the world. In China Windows XP may still be supported with software updates such as security patches.

Arduino Software IDE

The Arduino IDE is the program that is used to develop the program code that runs on and controls the Arduino. For example, in order for you to have the Arduino control the lighting state of an L.E.D. (Light Emitting Diode) you will need to write a computer program in C/C++ using the Arduino IDE. Then, you will need to compile this program into a form that the Arduino is able to execute and then transfer the final compiled program using the Arduino IDE. From there the program automatically executes and controls the L.E.D. that is connected to the Arduino.

New versions of the IDE are compiled daily or hourly and are available for download. Older versions of the IDE are also available for downloading at:

http://www.arduino.cc/en/Main/OldSoftwareReleases

In this section we will go over the key features of the Arduino Software IDE. The IDE you are using may be slightly different then the version discussed in this section but the general functions we cover here should still be the same. We won't go in depth into every detail of the IDE since this book is meant as a quick start guide and not a reference manual. We will cover the critical features of the Arduino IDE that you will need to get started on the projects in this book. See Figure 1-14.

Figure 1-14. The Arduino IDE

The "verify" button checks to see if the program you have entered into the Arduino IDE is valid and without errors. These programs are called "sketches". See Figure 1-15.

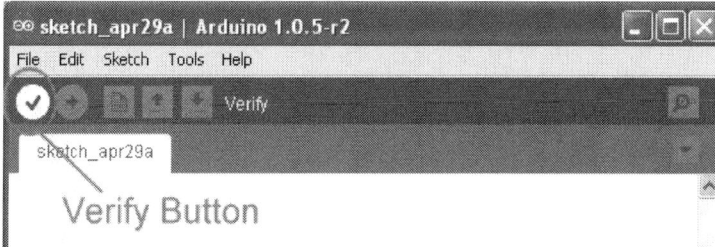

Figure 1-15. The verify button

The "upload" button first verifies that the program in the IDE is a valid C/C++ program with no errors, compiles the program into a form the Arduino can execute, and then finally transfers the program via the USB cable that is connected to your computer to your Arduino board. See Figure 1-16.

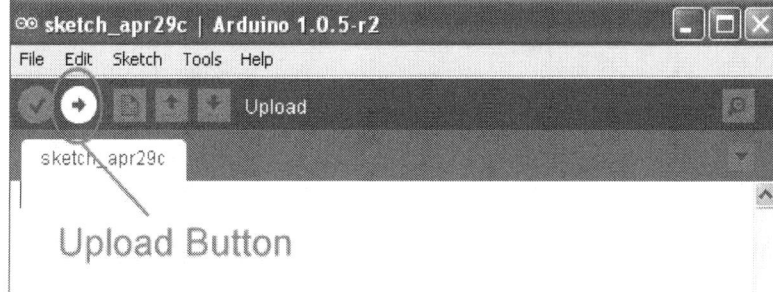

Figure 1-16. The upload button

The "New" button creates a new blank file or sketch inside the Arduino IDE where the user can create his or her own C/C++ program for verification, compilation, and transferring to the Arduino. See Figure 1-17.

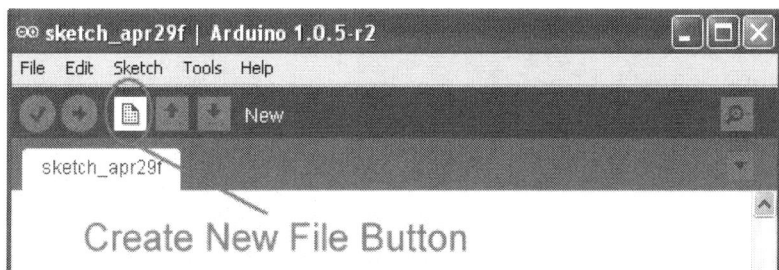

Figure 1-17. The new file button

The "open" file button is used to open and load in the Arduino C/C++ program source code from a file or load in various sample source codes from example Arduino projects that are included with the IDE. See Figure 1-18.

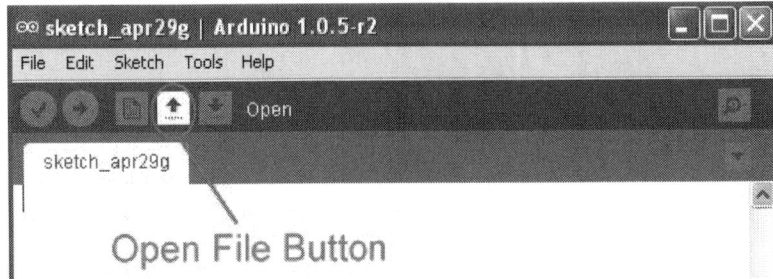

Figure 1-18. The open file button

The "save" button saves the sketch you are currently working on to disk. A file save dialog is brought up first and then you will able to save the file on your computer's hard drive. See Figure 1-19.

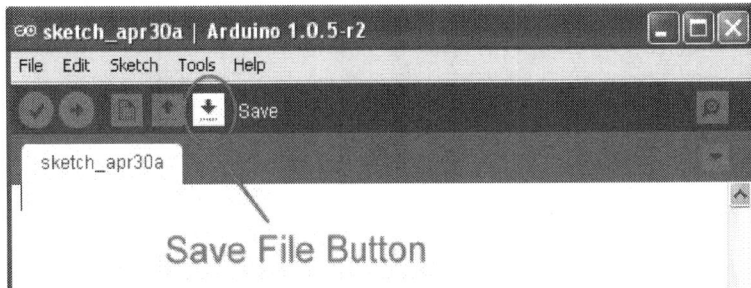

Figure 1-19. Save button

The "serial monitor" button brings up the Serial Monitor debug program where the user can examine the output of debug statements from the Arduino program. The Serial Monitor can also accept user input that can be processed by the Arduino program. We will discuss the Serial Monitor and how to use it as a debugger and input console later in this book. See Figure 1-20.

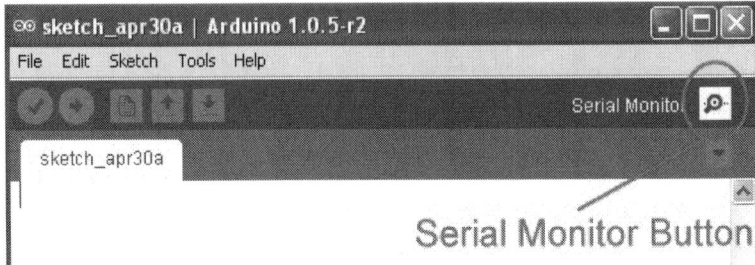

Figure 1-20. Serial Monitor

There are also other important features of the main window of the Arduino IDE. The title bar of the IDE window contains the Arduino IDE version number. In Figure 1-21 the Arduino version number is 1.0.5 r2. The sketch name is displayed in the source code tab and is "Blink" which is one of the sample sketches that come with the Arduino IDE. The source code area which is the large white area with scrollbars on the right side and bottom is where you enter your C/C++ source code that will control the behavior of the Arduino. The bottom black area in the IDE is where warning and errors are displayed from the code verification process. At the bottom left hand corner of the IDE is a number that represents the line number in the source code where the user's cursor is currently located. In the lower right hand corner of the IDE is the currently selected Arduino model and COM port that the Arduino is attached to.

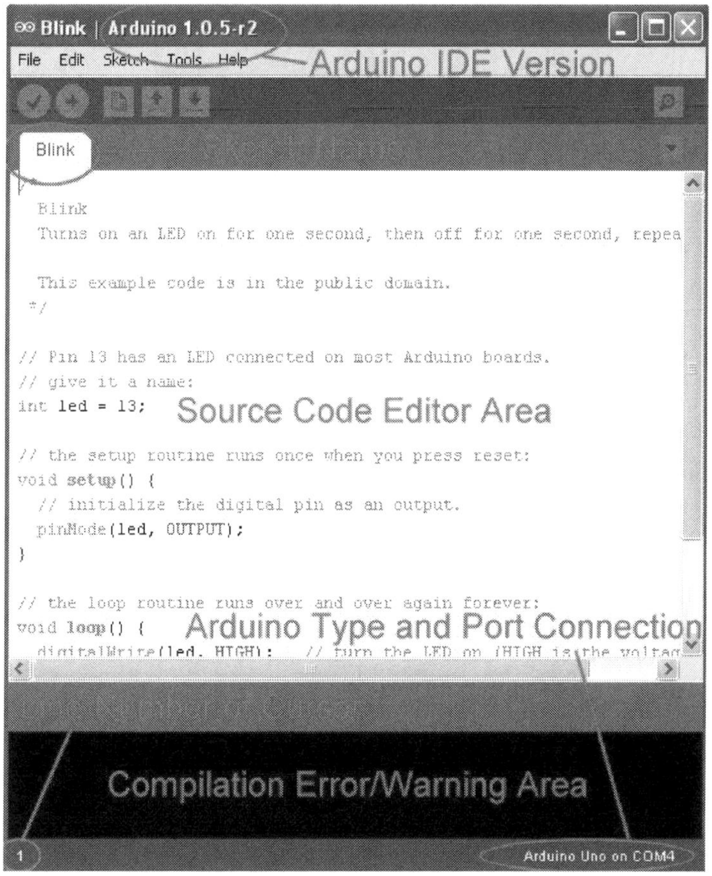

Figure 1-21. The General IDE

Hands on Example: A simple Arduino "Hello World" program with an LED

In this hands on example I show you how to set up the Arduino development system on your Windows based PC or Mac. I first discuss where you can get an Arduino board and USB cable. Then I discuss the installation of the Arduino IDE and Arduino hardware device drivers. I then discuss how to load in the "Blink" sketch example program. Next, I tell you how to verify that the program is without syntax errors, how to upload it onto the Arduino, and how to tell if the program is working. Finally, I discuss how the Blink program code works and show you how to modify it.

Get an Arduino Board and USB Cable

You can purchase an official Arduino Uno board from a distributor listed on http://www.arduino.cc/en/Main/Buy or http://www.Arduino.org. The first web site is still generally considered the main web site for Arduino. However, the second web site is run by the people actually making Arduino boards. The split among the founders of the Arduino mentioned earlier can be seen here in terms of who is designated as a distributor of an "official" Arduino board.

A second option is to buy an unofficial Arduino Uno made by a seller not listed as an "official" distributor by either arduino.cc or arduino.org. These boards are generally a lot cheaper than an "official" Arduino board. However, the quality may vary widely between manufacturers or even between production runs between the same manufacturer.

In terms of the USB cable that is used to connect the Arduino to your development computer the official Arduino board generally does not come with a cable but many unofficial boards come with short USB cables. Arduino compatible USB cables of a longer length such as 6 foot or 10 foot can be bought on Amazon.com or Ebay.com. I purchased the "Mediabridge USB 2.0 - A Male to B Male Cable (10 Feet) - High-Speed with Gold-Plated Connectors – Black" from Amazon.com for my "official" Arduino Uno and its seems to work well. Make sure you get the right kind of USB cable with the right connectors on either end. The rectangular end of the USB cable is connected to your computer and the square end is connected to your Arduino. See Figure 1-22.

Figure 1-22. Arduino USB Cable

Install the Arduino IDE

The Arduino IDE has versions that can run on the Windows, Mac, and Linux operating systems. The Arduino IDE can be downloaded from:

http://www.arduino.cc/en/Main/Software

I recommend installing the Windows executable version if you have a windows based computer. Follow the directions in the pop windows.

> Note: The Arduino web site also contains links to instructions for installing the Arduino IDE for Windows, Mac, and Linux on http://www.arduino.cc/en/Guide/HomePage. The installation for Linux depends on the exact version of Linux being used.

Install the Arduino Drivers

The next step is to connect your Arduino to your computer using the USB cable. If you are using Windows it will try to automatically install your new Arduino hardware. Follow the directions in the pop up windows to install the drivers. Decline to connect to Windows Update to search for the driver. Select "Install the software automatically" as recommended. If you are using XP ignore the popup window warning about the driver not passing windows logo testing to verify its compatibility with XP.

If this does not work then instead of selecting "Install the software automatically" specify a specific driver location which is the "drivers/FTDI" directory under your main Arduino installation directory.

Loading in the Blink Arduino Sketch Example

Next, we need to load the Blink sketch example into the Arduino IDE. Click the "open" button to bring up the menu. Under "01. Basics" select the "Blink" example to load in. See Figure 1-23.

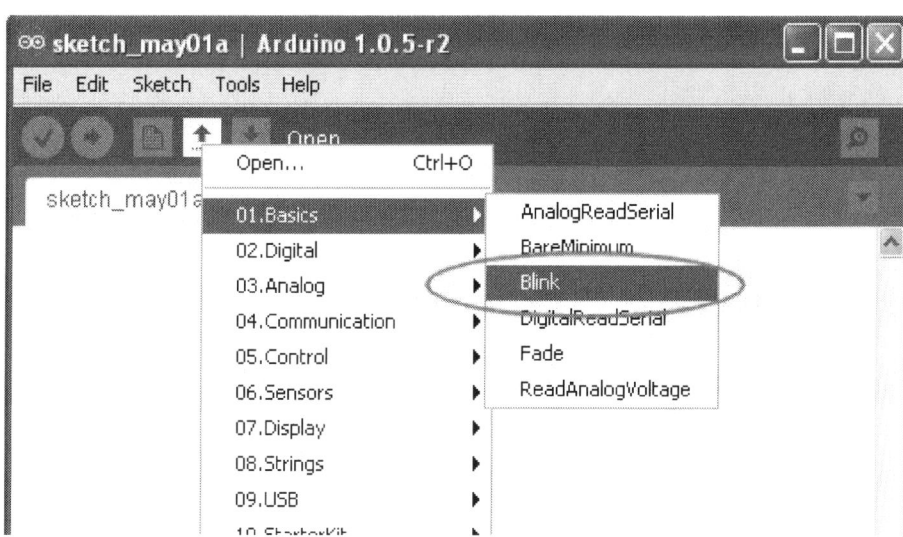

Figure 1-23. Loading in the blink example

The code that is loaded into the Arduino IDE should look like the code in Listing 1-1.

Listing 1-1. Blink Sketch

/*

Blink

Turns on an LED on for one second, then off for one second, repeatedly.

This example code is in the public domain.

*/

// Pin 13 has an LED connected on most Arduino boards.

// give it a name:

```
int led = 13;


// the setup routine runs once when you press reset:
void setup() {
  // initialize the digital pin as an output.
  pinMode(led, OUTPUT);
}


// the loop routine runs over and over again forever:
void loop() {
  digitalWrite(led, HIGH);   // turn the LED on (HIGH is the voltage level)
  delay(1000);               // wait for a second
  digitalWrite(led, LOW);    // turn the LED off by making the voltage LOW
  delay(1000);               // wait for a second
}
```

Verifying the Blink Arduino Sketch Example

Click the "verify" button to verify the program is valid C/C++ code and is error free.   See Figure 1-24.

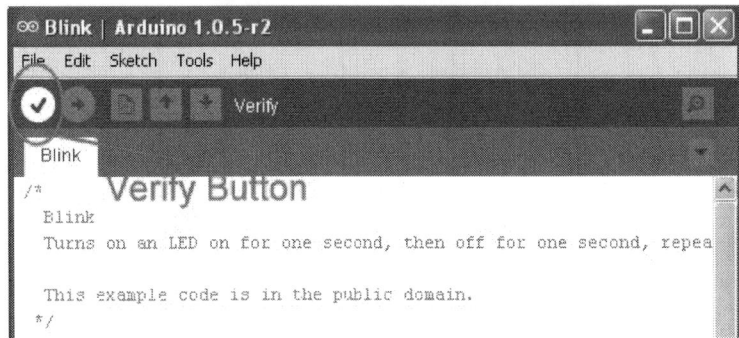

Figure 1-24. Verifying the blink sketch

Uploading the Blink Arduino Sketch Example

Before uploading the sketch to your Arduino make sure the type of Arduino under the "Tools->Board" menu item is correct. In our case the board type should be set to be "Arduino Uno". See Figure 1-25.

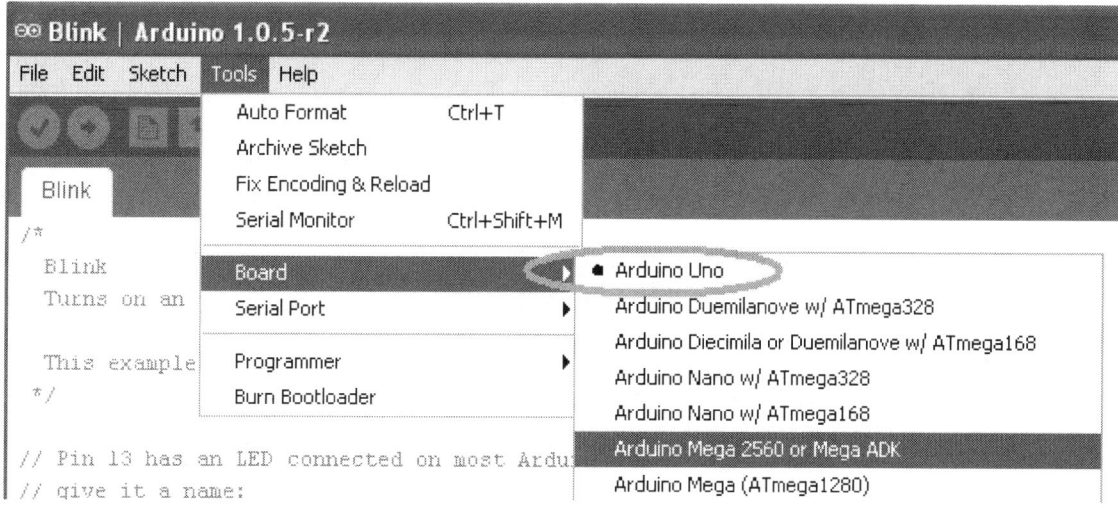

Figure 1-25. Set Arduino type to Arduino Uno

Next, make sure the serial port is set correctly to the one that is being used by your Arduino. Generally Com1, and Com2 are reserved and the serial port that the Arduino will be connected to is Com3 or higher. See Figure 1-26.

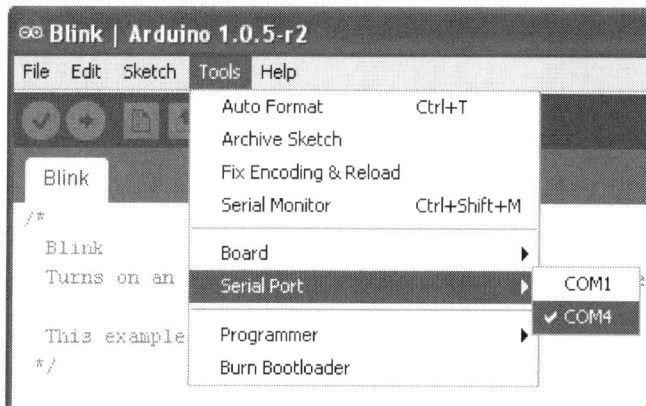

Figure 1-26. Set the com port

If you are using a Mac then the Serial Port selection should be something like "/dev/tty.usbmodem" instead of a COMXX value.

Next, with the Arduino connected press the "upload" button to verify, compile, and then transfer the Blink example program to the Arduino. After the program has finished uploading you should see a message that the upload has been completed in the warnings/error window at the bottom of the IDE inside the black window. See Figure 1-27.

24

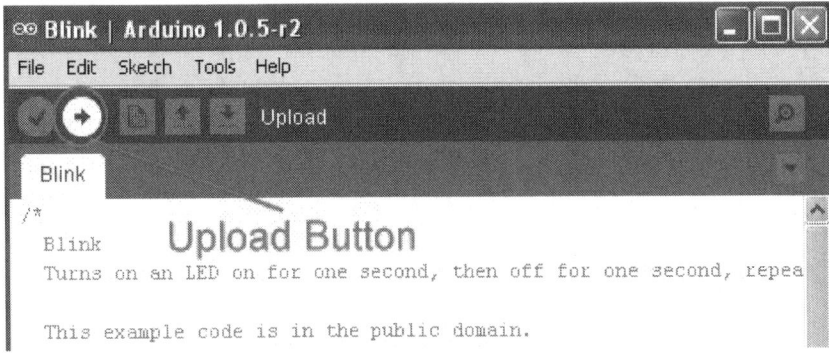

Figure 1-27. Upload to Arduino

> Note: The Upload button does the job of the "verify" button but also uploads the final compiled program to the Arduino.

The Final Result

The final result will be a blinking light on the Arduino board near digital pin 13. By design the Arduino board has a built in L.E.D. connected to pin 13. So this example did not require you to connect an actual separate L.E.D. to the Arduino board. See Figure 1-28.

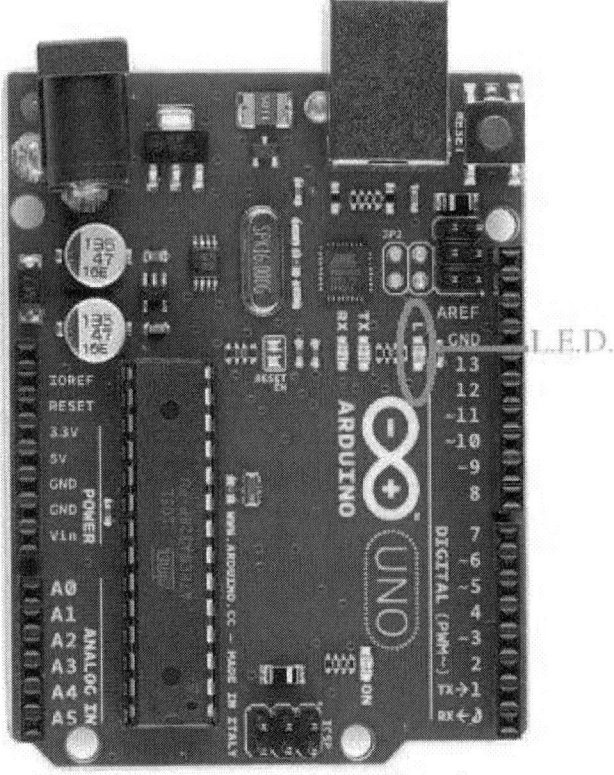

Figure 1-28. Built in L.E.D.

Playing Around with the Code

The default of the program is to turn on the L.E.D. for one second and then turn off the L.E.D. for one second. The code that controls the timing is located in the loop() function. The digitalWrite() function sets the variable led which is pin 13 to either on which is HIGH or off which is LOW. The delay() function suspends the execution of the program for 1000 milliseconds or 1 second so that the L.E.D. is set on for 1 second and off for 1 second. See Listing 1-2.

Listing 1-2. loop() Function

```
void loop() {

    digitalWrite(led, HIGH);   // turn the LED on (HIGH is the voltage level)

    delay(1000);               // wait for a second

    digitalWrite(led, LOW);    // turn the LED off by making the voltage LOW

    delay(1000);               // wait for a second

}
```

Play around with the values in the delay() functions lengthening or shortening the time the L.E.D. stays on and/or lengthening or shortening the time the L.E.D. stays off. For example, to have the L.E.D. briefly flash shorten the first delay value to 100. This would shorten the time that the L.E.D. stays on. Upload the new sketch to the Arduino by pressing the "upload" button. After it has finished uploading you see a message indicating that in the black warnings/error message window at the bottom of the IDE. Look at the L.E.D. on the Arduino. The timing of the L.E.D. on/off pattern should have changed.

Summary

In this chapter I introduced the Arduino to the reader. I concentrated my coverage on the Arduino Uno. I discussed the UNO's basic features and then covered key functional components. Next, information about the software needed to develop programs for the Arduino called the Arduino IDE (Integrated Development Environment) was presented including the different versions of the IDE available for different platforms. Information about key features of the IDE was discussed. Finally, a "Hands on Example" was presented where I took the reader through a step by step guide to setting up the Arduino with a development computer system. In this example I also discussed loading in an example program, uploading this example program to the Arduino, and then modifying the program to see how this changes the output.

Chapter 2

Arduino Programming Language Basics

In this chapter I go over the basics of the Arduino Language. I cover the basic elements of the Arduino language that you will need in order to create programs that control the Arduino board. Various key elements such as data types, constants, control loops, etc. are covered.

C/C++ Language for Arduino Overview

The Arduino uses C and C++ in its programs called "sketches". This section briefly summaries key language elements. This is not meant as a reference guide and ideally you should have some experience with a programming language similar to C and/or C++.

Comments

- // - This is a single line comment that is used by the programmer to document the code. These comments are not executed by the Arduino.

- /* */ - These enclose a multi line comment that is used by the programmer to document the code. These comments are not executed by the Arduino.

Data Types

- void – This type being used with a function indicates that the function will not return any value to the function caller. For example, the setup() function that is part of the standard Arduino code framework has a return type of void.

    ```
    void setup()
    {
        // Initialize the Arduino, camera, and SD card here
    }
    ```

- boolean – A boolean variable can hold either the value of true or false and is 1 byte in length. For example, in the following code the variable result is declared of type boolean and is initialized to false.

boolean result = false;

- char – The char variable type can store character values and is 1 byte in length. The following code declares that tempchar is of type char and is an array with 50 elements.

  char tempchar[50];

- unsigned char – The unsigned char data type holds 1 byte of information in the range of 0 through 255.

- byte – The byte data type is the same as the unsigned char data type. The following code declares a variable called data of type byte that is initialized to 0.

  byte data = 0;

- int – The int data type holds a 2 byte number in the range of -32,768 to 32,767.

- unsigned int – This data type is 2 bytes in length and holds a value from 0 to 65,535.

- word - This data type is the same as the unsigned int type.

- long – This data type is 4 bytes in length and holds a value from -2,147,483,648 to 2,147,483,647.

- unsigned long – This data type is 4 bytes in length and holds a value between 0 to 4,294,967,295.

- Float – This is a floating point number that is 4 bytes in length and holds a value between -3.4028235E+38 to 3.4028235E+38.

- double - On the current Arduino implementation the double is the same as float with no gain in precision.

- String – This is a class object that allows the user to easily manipulate groups of characters. In the following code a new variable called Command of type String is declared and initialized to the QQVGA resolution.

  String Command = "QQVGA";

- array – An array is a continuous collection of data that can be accessed by an index number. Arrays are 0 based so that the first element in the array has an index of 0. Common types of arrays are character arrays, and integer arrays. The following code declares the variable Entries as an array of type String that contains 10 elements. The function ProcessRawCommandElement() is then called with element number 2 in the Entries array which is the third element in the array. Remember 0 is the first element in the array.

  String Entries[10];

  boolean success = ProcessRawCommandElement(Entries[2]);

Constants

- INPUT – This is an Arduino pin configuration that sets the pin as an input pin that allows you to easily read in the voltage value at that pin with respect to ground on the Arduino. For example, the following code sets the pin VSYNC on the Arduino as an INPUT pin which allows you to read in the voltage value of the pin. The function pinMode() is an Arduino function included in the built in library.

    pinMode(VSYNC, INPUT);

- OUTPUT – This is an Arduino pin configuration that sets the pin as an output pin that allows you to drive other electronics components such as an L.E.D. or to provide digital input to other devices in terms of HIGH or LOW voltages. In the following code the pin WEN is set to OUTPUT using the built in pinMode() function.

    pinMode(WEN , OUTPUT);

- HIGH (pin declared as INPUT) - If a pin on the Arduino is declared as an INPUT then when the digitalRead() function is called to read the value at that pin then a HIGH value would indicate a value of 3 volts or more at that pin.

- HIGH (pin declared as OUTPUT) – If a pin on the Arduino is declared as an OUTPUT then when the pin is set to HIGH with the digitalWrite() function then the pin's value is 5 volts.

- LOW (pin declared as INPUT)- If a pin on the Arduino is declared as an INPUT then when the digitalRead() function is called to read the value at that pin then a LOW value would indicate a value of 2 volts or less.

- LOW (pin declared as OUTPUT) - If a pin on the Arduino is declared as an OUTPUT then when the digitalWrite() function is called to set the pin to LOW the voltage value at that pin would be set to 0 volts.

- true – true is defined as any non zero number such as 1, -1, 200, 5, etc.

- false – false is defined as 0.

The Define Statement

The define statement assigns a name to a constant value. During the compilation process the compiler will replace the constant name with the constant value.

#define constantName value

The following defines the software serial data receive pin as pin 6 on the Arduino and the software serial data transmit pin as pin 7 on the Arduino.

#define RxD 6

#define TxD 7

These definitions are used in defining which pins are to receive and transmit data via the software serial method which is initialized as shown below and are used to communicate with the Bluetooth adapter that will be connected to the Arduino in the examples in this book.

SoftwareSerial BT(RxD,TxD);

The Include Statement

The #include statement brings in code from outside files and "includes" them into your Arduino sketch. Generally a header or .h file is included which allows access to the functions and classes inside that file.

For example, in a program we can include the Wire.h file which let's us use the Wire library. The Wire library has functions to initialize, to read data from and to write data to a device connected to the I2C interface. We need the Wire library to use a device that uses the I2C bus.

#include <Wire.h>

The Semicolon

Each statement in C/C++ needs to end in a semicolon. For example, when declaring and initializing a variable you will need a semicolon.

const int chipSelect = 48;

When you use a library that you included with the #include statement you will need a semicolon at the end when you call a function.

Wire.begin();

Curly Braces

The curly braces such as { and } specify blocks of code and must match in pairs. That is, for every opening brace { there must be a closing brace } to match.

A function requires curly braces to denote the beginning and end of the function.

void Function1()

{

  // Body of Function

}

Program loops such as the for statement may also need the curly braces

for (int I = 0; I < 9; I++)

{

```
    // Body of loop
}
```

It is also good practice to use braces in control structures such as the if statement.

```
if (I < 0)
{
    // Body of If statement
}
```

Arithmetic Operators

- = - The equals sign is the assignment operator used to set a variable to a value. For example, the following sets the value of the variable Data to the result from the function CreatePhotoInfo().

  String Data = CreatePhotoInfo();

- + - The plus sign performs addition. For example, the following adds the Strings Command, PhotoTakenCount and Ext together to get a final string called Filename.

  String Filename = Command + PhotoTakenCount + Ext;

- - - The minus sign performs subtraction. For example, the following calculates the time it takes to capture a photo using the camera by measuring the difference between the starting time before the image is captured and the ending time just after the image is captured.

  ElapsedTime = TimeForCaptureEnd - TimeForCaptureStart;

- * - The asterisk sign performs multiplication. For example, the total bytes of an image is calculated by multiplying the width of the image by the height of the image by the bytes per pixel in the image.

  int TotalBytes = ImageWidth * ImageHeight * BytesPerPixel;

- / - The back slash sign performs division. For example, the speed in miles per hour of an object is calculated by dividing the number of miles the object has traveled by the number of hours that it took to travel that distance.

  float Speed = NumberMiles / NumberHours;

- % - The percent sign is the modulo operator that returns the remainder from a division between two integers. For example,

  int remainder = dividend % divisor;

Comparison Operators

- == - The double equal is a comparison operator to test if the argument on the left side of the double equal sign is equal to the argument on the right side. If the arguments are equal then it evaluates to true. Otherwise it evaluates to false. For example, if Command is equal to "SystemStart" then the code block is executed.

    ```
    if (Command == "SystemStart")
    {
            // Execute code
    }
    ```

- != - The exclamation point followed by an equal sign is the not equal to operator that evaluates to true if the argument on the left is not equal to the argument on the right side. Otherwise, it evaluates to false. For example, in the following code if the current camera resolution is not set to VGA then the code block is executed.

    ```
    if (Resolution != VGA)
    {
            // If current resolution is not VGA then set camera for VGA
    }
    ```

- < - The less than operator evaluates to true if the argument on the left is less than the argument on the right. For example, in the code that follows the for loop will execute the code block while the height is less than the height of the photo. When the height counter becomes equal or greater than the photo's height then the loop exits.

    ```
    for (int height = 0; height < PHOTO_HEIGHT; height++)
    {
            // Process every row of the photo
    }
    ```

- > - The greater than operator evaluates to true if the argument on the left side is greater than the argument on the right side. For example, in the following code if there are available characters to read in from the Serial Monitor then execute the code block. That is the number of available characters to read in must be greater than 0.

    ```
    if (Serial.available() > 0)
    {
            // Process available characters from Serial port
    ```

- <= - The less than sign followed by the equal sign returns true if the argument on the left hand side is less than or equal to the argument on the right hand side. It returns false otherwise.

- \>= - The greater than sign followed by the equal sign returns true if the argument on the left is greater than or equal to the argument on the right. It returns false otherwise.

Boolean Operators

- && - This is the "and" boolean operator that only returns true if both the arguments on the left and right side evaluate to true. It returns false otherwise. For example, in the following code only if the previous value is 'O' and the current value is 'K' will the code block be executed. Otherwise it will not be executed.

    ```
    char out,outprev = '$';

    if ((outprev == 'O')&&(out == 'K'))

    {

        out = ReadinData();

        // Code block

        outprev = out;

    }
    ```

- || - This is the "or" operator and returns true if either the left argument or the right argument evaluates to true. Otherwise, it returns false. For example, in the following code if either the camera's command is set to QQVGA or QVGA then the code block is executed. Otherwise it is not executed.

    ```
    if ((Command == "QQVGA") || (Command == "QVGA"))

    {

        // Code

    }
    ```

- ! – The not operator returns the opposite boolean value. The not value of true is false which is 0 and the not value of false is true which is non zero. In the following code a file is opened on the SD card and a pointer to the file is returned. If the pointer to the file is NULL which has a 0 value then not NULL would be 1 which is true. The if statement is executed when the argument is evaluated to true which means that the file pointer is NULL. This means that the open operation has failed and an error message needs to be displayed.

    ```
    // Open File
    ```

```
InfoFile = SD.open(Filename.c_str(), FILE_WRITE);

// Test if file actually open

if (!InfoFile)

{

    Serial.println(F("\nCritical ERROR ... Can not open Photo Info File for output ... "));

    return;

}
```

Bitwise Operators

- & - This is the bitwise "and" operator between two numbers where each bit of each number has the "and" operation performed on it to produce the result in the final number. The resulting bit is 1 only if both bits in each number is 1. Otherwise the resulting bit is 0.

- | - This is the bitwise "or" operator between two numbers where each bit of each number has the "or" operation performed on it to produce the result in the final number. The resulting bit is 1 if the bit in either number is 1. Otherwise the resulting bit is 0.

- ^ - This is the bitwise "xor" operator between two numbers where each bit of each number has the "exclusive or" operation performed on it to produce the result in the final number. The resulting bit is 1 if the bits in each number are different and 0 otherwise.

- ~ - This is the bitwise "not" operator where each bit in the number following the "not" symbol is inverted. The resulting bit is 1 if the initial bit was 0 and the bit is 0 if the initial bit was 1.

- << - This is the bitshift left operator where each bit in the left operand is shifted to the left by the number of positions indicated by the right operand. For example, in the code below a 1 is shifted left PinPosition times and the final value is assigned to the variable ByteValue.

    ByteValue = 1 << PinPosition;

- >> - This is the bitshift right operator where each bit in the left operand is shifted to the right by the number of positions indicated by the right operand. For example, in the code below bits in the number 255 are shifted to the right PinPosition times and the final value is assigned to the variable ByteValue.

    ByteValue = 255 >> PinPosition;

Compound Operators

- ++ - This is the increment operator. The exact behavior of this operator also depends if it is placed before the variable being incremented or after the variable being incremented. In the following code the variable PhotoTakenCount is incremented by 1.

    PhotoTakenCount++;

    If the increment operator is placed after the variable being incremented then the variable is used first in the expression it is in before being incremented. For example, in the code below the height variable is used first in the for loop expression before it is incremented. So the first iteration of the for loop below would use height = 0. The height variable would be incremented after being used in the expression.

    for (int height = 0; height < PHOTO_HEIGHT; height++)

    {

        // Process row of image

    }

    If the increment operator is placed before variable being incremented then the variable is incremented first then it is used in the expression that it is in. For example, in the code below the height variable is incremented first before it is used in the for loop. This means that in the first iteration of the loop the height variable is 1 instead of 0.

    for (int height = 0; height < PHOTO_HEIGHT; ++height)

    {

        // Process row of image

    }

- -- - The decrement operator decrements a variable by 1 and its exact behavior depends on the placement of the operator either before or after the variable being decremented. If the operator is placed before the variable then the variable is decremented before being used in an expression. If the operator is placed after the variable then the variable is used in an expression before it is decremented. This follows the same pattern as the increment operator discussed previously.

- += - The compound addition operator adds the right operand to the left operand. This is actually a short hand version of

    operand1 = operand1 + operand2;

    Which is the same as the version that uses the compound addition operator.

    operand1 += operand2;

- -= - The compound subtraction operator subtracts the operand on the right from the operand on the left. For example, the code for a compound subtraction would be:

operand1 -= operand2;

This is the same as the following:

operand1 = operand1 - operand2;

- *= - The compound multiplication operator multiplies the operand on the right by the operand on the left. The code for this is as follows.

    operand1 *= operand2;

    This is also equivalent to:

    operand1 = operand1 * operand2;

- /= - The compound division operator divides the operand on the left by the operand on the right. For example,

    operand1 /= operand2;

    This is equivalent to:

    operand1 = operand1 / operand2;

- &= - The compound bitwise and operator is equivalent to:

    x = x & y;

- != - The compound bitwise or operator is equivalent to:
    - x = x | y;

Pointer Access Operators

- * - The de-reference Operator allows you to access the contents that a pointer points to. For example, the code that follows declares a variable called pdata as a pointer to a byte and creates storage for the data using the new command. The pointer variable called pdata is then de-referenced to allow the actual data that the pointer points to be set to 1.

    byte *pdata = new byte;

    *pdata = 1;

- & - The address operator creates a pointer to a variable. For example, the following code declares a variable called data of type byte and assigns the value of 1 to it. A function called FunctionPointer() is defined that accepts as a parameter a pointer to a byte. In order to use this function with the variable data we need to call that function with a pointer to the variable data.

    byte data = 1;

    void FunctionPointer(byte *data)

    {

```
    // body of function

}

FunctionPointer(&data);
```

Variable Scope

- Global variables – In the Arduino programming environment global variables are variables that are declared outside any function and before they are used.

    ```
    // VGA Default

    int PHOTO_WIDTH  =  640;

    int PHOTO_HEIGHT =  480;

    int PHOTO_BYTES_PER_PIXEL = 2;
    ```

- Local variables – Local variables are declared inside functions or code blocks and are only valid inside that function or code block. For example, in the following function the variable localnumber is only visible inside the Function1() function.

    ```
    void Function1()
    {
        int localnumber = 0;
    }
    ```

Conversion

- char(x) – This function converts a value x into a char data type and then returns it.

- byte(x) – This function converts a value x into a byte data type and then returns it.

- int(x) – This function converts a value x into a integer data type and then returns it.

- word(x) – This function converts a value x into a word data type and then returns it.

- word(highbyte,lowbyte) – This function combines two bytes, the high order byte and the low order byte into a single word and then returns it.

- long(x) – This function converts a value x into a long and then returns it.

- float(x) – This function converts a value x into a float and then returns it.

Control Structures

- if (comparison operator) – The if statement is a control statement that tests if the result of the comparison operator or argument is true. If it is true then execute the code block. For example, in the following code the if statement tests to see if there is more data from the Bluetooth connection that needs to be read in. If there is then read the data in and assign it to the RawCommandLine variable.

    if(BT.available() > 0)

    {

    // If command is coming in then read it

    RawCommandLine = GetCommand();

    break;

    }

- if (comparison operator) else – The if else control statement is similar to the if statement except with the addition of the else section which is executed if the previous if statement evaluates to false and is not executed. For example, in the following code if the frames per second parameter is set for 30 frames per second then the SetupCameraFor30FPS() function is called. Otherwise, if the frames per second parameter is set to night mode then the SetupCameraNightMode() function is called.

    // Set FPS for Camera

    if (FPSParam == "ThirtyFPS")

    {

      SetupCameraFor30FPS();

    }

    else

    if (FPSParam == "NightMode")

    {

      SetupCameraNightMode();

    }

- for (initialization; condition; increment) – The for statement is used to execute a code block usually initializing a counter then performing actions on a group of objects indexed by that incremented value. See the code example below.

    for (int i = 0 ; i < NumberElements; i++)

    {

```
            // Process Element i  Here

    }
```

- while (expression) – The while statement executes a code block repeatedly until the expression evaluates to false. In the following code the while code block is executed as long as data is available for reading from the file.

    ```
    // read from the file until there's nothing else in it:

    while (TempFile.available())

    {

        Serial.write(TempFile.read());

    }
    ```

- break – A break statement is used to exit from a loop such as a while or for loop. In the following code the while loop causes the code block to be executed forever. If there is data available from the Serial Monitor then it is processed and then the while loop is exited.

    ```
    while (1)

    {

            if (Serial.available() > 0)

            {

                    // Process the data

                    break;

            }

    }
    ```

- return (value) – The return statement exits a function. It also may return a value to the calling function.

    ```
    return;

    return false;
    ```

Object Oriented Programming

The Arduino programming environment also supports the object oriented programming aspects of C++. An example of Arduino code that uses object oriented programming is the ArduCAM class that was developed to support the ArduCAM Mini camera. A C++ class is composed of data, functions that use that data, and a constructor that is used to create an object of the class.

An example of class data is:

byte sensor_model;

An example of a constructor which takes the camera model number and the chip select pin as input parameters is:

ArduCAM(byte model,int CS);

An example of a class function is:

void start_capture(void);

See Listing 2-1 for the full class declaration.

Listing 2-1. The ArduCAM class

```
class ArduCAM
{
    public:
        ArduCAM();
        ArduCAM(byte model,int CS);
        void InitCAM();

        void CS_HIGH(void);
        void CS_LOW(void);

        void flush_fifo(void);
        void start_capture(void);
        void clear_fifo_flag(void);
        uint8_t read_fifo(void);

        uint8_t read_reg(uint8_t addr);
        void write_reg(uint8_t addr, uint8_t data);

        uint32_t read_fifo_length(void);
        void set_fifo_burst(void);
```

```cpp
        void set_bit(uint8_t addr, uint8_t bit);
        void clear_bit(uint8_t addr, uint8_t bit);
        uint8_t get_bit(uint8_t addr, uint8_t bit);
        void set_mode(uint8_t mode);


        int wrSensorRegs(const struct sensor_reg*);
        int wrSensorRegs8_8(const struct sensor_reg*);
        int wrSensorRegs8_16(const struct sensor_reg*);
        int wrSensorRegs16_8(const struct sensor_reg*);
        int wrSensorRegs16_16(const struct sensor_reg*);


        byte wrSensorReg(int regID, int regDat);
        byte wrSensorReg8_8(int regID, int regDat);
        byte wrSensorReg8_16(int regID, int regDat);
        byte wrSensorReg16_8(int regID, int regDat);
        byte wrSensorReg16_16(int regID, int regDat);


        byte rdSensorReg8_8(uint8_t regID, uint8_t* regDat);
        byte rdSensorReg16_8(uint16_t regID, uint8_t* regDat);
        byte rdSensorReg8_16(uint8_t regID, uint16_t* regDat);
        byte rdSensorReg16_16(uint16_t regID, uint16_t* regDat);


        void OV2640_set_JPEG_size(uint8_t size);
        void OV5642_set_JPEG_size(uint8_t size);
        void set_format(byte fmt);


        int bus_write(int address, int value);
        uint8_t bus_read(int address);
    protected:
```

```
        regtype *P_CS;

        regsize B_CS;

        byte m_fmt;

        byte sensor_model;

        byte sensor_addr;
};
```

Summary

In this chapter I covered the basics of the Arduino programming language. I covered a broad range of basic topics such as data types, constants, built in functions, and control loops. In addition, this chapter was not meant to be a reference manual but a quick start guide to the basics of the Arduino programming language. Please refer to the official Arduino language reference for more information.

Chapter 3

The Android Controller and Bluetooth Communication with Arduino

In this chapter I give an overview of Android development and Bluetooth communication. I discuss the various development software for Android such as the Android Studio and the Android ADT bundle and what hardware and software is required to develop applications for Android devices. Then I give a basic review of the Java programming language for Android. Next, I discuss how I implemented Bluetooth communication on Android. Finally, I discuss how I implemented Bluetooth communication on the Arduino.

What is an Android?

The Android is a mobile operating system for cell phones and tablets. Android equipped devices are generally low cost and widely available which makes them the ideal device to use in a wireless home security system. Additionally many Android phones can be operated based on a month to month basis with no long term contract to sign. This is also ideal for an on demand type security system where you can scale up or down based on the need. For example, if the area where you live has experienced a series of thefts you can set up the burglar alarm security system described in this book. You can then activate the cell phone service for your Android phone so that if an intruder breaks into your home you can have the Android set to call out to an emergency number that you can designate to alert you of the possible break in. You can hide your Android in a location near areas that burglars might be interested in such as a home theater system and listen in to see if you hear any unusual activity. In this case you would probably hear the thieves moving the electronic equipment around so this would confirm that there is an actual break in. You would then call the police and report the crime as a burglary in progress.

Getting Started with Android Development

In this section we cover the different development environments available for Android development. First I cover the Android Studio development environment which is the newest. The Eclipse with ADT plug-ins is covered next along with the Eclipse ADT Bundle which is the preferred version of the Eclipse with ADT plug-ins.

Android Studio

Android Studio is the newest Android development system available and is recommended by Google.

System Requirements:

Windows

- Microsoft® Windows® 8/7/Vista (32 or 64-bit)
- 2 GB RAM minimum, 4 GB RAM recommended
- 400 MB hard disk space
- At least 1 GB for Android SDK, emulator system images, and caches
- 1280 x 800 minimum screen resolution
- Java Development Kit (JDK) 7
- Optional for accelerated emulator: Intel® processor with support for Intel® VT-x, Intel® EM64T (Intel® 64), and Execute Disable (XD) Bit functionality

Mac OS X

- Mac® OS X® 10.8.5 or higher, up to 10.9 (Mavericks)
- 2 GB RAM minimum, 4 GB RAM recommended
- 400 MB hard disk space
- At least 1 GB for Android SDK, emulator system images, and caches
- 1280 x 800 minimum screen resolution
- Java Runtime Environment (JRE) 6
- Java Development Kit (JDK) 7
- Optional for accelerated emulator: Intel® processor with support for Intel® VT-x, Intel® EM64T (Intel® 64), and Execute Disable (XD) Bit functionality
- On Mac OS, run Android Studio with Java Runtime Environment (JRE) 6 for optimized font rendering. You can then configure your project to use Java Development Kit (JDK) 6 or JDK 7.

Linux

- GNOME or KDE desktop

- GNU C Library (glibc) 2.15 or later

- 2 GB RAM minimum, 4 GB RAM recommended

- 400 MB hard disk space

- At least 1 GB for Android SDK, emulator system images, and caches

- 1280 x 800 minimum screen resolution

- Oracle® Java Development Kit (JDK) 7

- Tested on Ubuntu® 14.04, Trusty Tahr (64-bit distribution capable of running 32-bit applications).

> Note: The official web page for Android Studio is http://developer.android.com/tools/studio/index.html

Eclipse with Android Development Tools (ADT)

All the Android project examples for this book were created in the Eclipse development environment using the Android Development Tools (ADT) plug-ins. This has an advantage that it works with older operating systems such as Windows XP. In fact inexpensive Windows XP systems can be bought on sites such as Ebay and Amazon.com. If you are starting from scratch and want an affordable system to develop Android applications then you should consider getting a Windows XP based computer system for Android as well as Arduino development.

Android development using the Eclipse with Android Development Tools plug-ins can be done on a Windows PC, Mac OS machine, or a Linux machine. The exact operating system requirements are as follows:

Operating Systems:

- Windows XP (32-bit), Vista (32- or 64-bit), or Windows 7 (32- or 64-bit)

- Mac OS X 10.5.8 or later (x86 only)

- Linux (tested on Ubuntu Linux, Lucid Lynx)

- GNU C Library (glibc) 2.7 or later is required.

- On Ubuntu Linux, version 8.04 or later is required.

- 64-bit distributions must be capable of running 32-bit applications.

Developing Android programs also requires installation of the Java Development Kit. Java Development Kit requirements are JDK 6 or later and are located at www.oracle.com/technetwork/java/javase/downloads/index.html.

If you are using a Mac, then Java may already be installed.

The Eclipse IDE program modified with the Android Development Tools (ADT) plug-in forms the basis for the Android development environment. The requirements for Eclipse are as follows:

- Eclipse 3.6.2 (Helios) or greater located at http://eclipse.org

- Eclipse JDT plug-in (included in most Eclipse IDE packages)

- Android Development Tools (ADT) plug-in for Eclipse located at http://developer.android.com/tools/sdk/eclipse-adt.html

> Note: Eclipse 3.5 (Galileo) is no longer supported with the latest version of ADT. For the latest information on Android development tools, go to http://developer.android.com/tools/index.html.
>
> The official web page for Android Eclipse Android Development Tools (ADT) plug ins was on http://developer.android.com/tools/help/adt.html

Eclipse ADT Bundle Download

If you are going to use the Eclipse with ADT plug-ins for your development then I suggest downloading the ADT Bundle which is basically a zip file that consists of a version of Eclipse with the ADT plug-ins already installed. All you would need to do is uncompress the zip file and create a shortcut to the "eclipse.exe" file that is in one of the directories inside the bundle.

The ADT Bundle is available for the Windows, Mac, and Linux operating systems at the following links that were working as this sentence is being typed. You can also search for "android adt bundle" using a search engine like Google or Yahoo to get the latest results.

Win 32 Bit:

https://dl.google.com/android/adt/adt-bundle-windows-x86-20140702.zip

Win 64 Bit:

https://dl.google.com/android/adt/adt-bundle-windows-x86_64-20140702.zip

Mac 64 Bit:

http://dl.google.com/android/adt/adt-bundle-mac-x86_64-20140702.zip

Linux 32 Bit:

http://dl.google.com/android/adt/adt-bundle-linux-x86-20140702.zip

Linux 64 Bit:

http://dl.google.com/android/adt/adt-bundle-linux-x86_64-20140702.zip

Migrating To Android Studio from Eclipse ADT

The Android Studio is the new "official" development environment for Android programs and replaces the Eclipse ADT system which will no longer be supported in terms of support for new versions of the Android operating system. However, for our purposes the older Eclipse ADT is fine and all the Android projects in this book were creating using the older Eclipse ADT bundle.

To migrate existing Android projects, simply import them using Android Studio:

1. In Android Studio, close any projects currently open. You should see the Welcome to Android Studio window.

2. Click Import Non-Android Studio project.

3. Locate the project you exported from Eclipse, expand it, select the build.gradle file and click OK.

4. In the following dialog, leave the "Use gradle wrapper" selected and click "OK". (You do not need to specify the Gradle home.)

5. Android Studio properly updates the project structure and creates the appropriate Gradle build file.

For a more full summary and additional information about migrating your Eclipse ADT project to the newest version of Android Studio visit the official web page for migration as shown below in the following important note.

> Note: The official web page for migrating code from the Android Eclipse ADT to the new Android Studio was on http://developer.android.com/sdk/installing/migrate.html

Overview of the Java Language

This section on the Java language is intended as a quick start guide to someone who has some knowledge of computer programming as well as some knowledge about object oriented programming. This section is NOT intended to be a Java reference manual. It is also not intended to cover every feature of the Java programming language.

The Java language for Android is run on a Java virtual machine. This means that the same compiled Java Android program can run on many different Android phones with different central processing unit (CPU) types. This is a key feature in terms of future expandability to faster processing units including those that will be specifically designed to enhance 3d games. The trade off to this is speed. Java programs run slower than programs compiled for a CPU in its native machine language since a Java virtual machine must interpret the code and then execute it on the native processor. A program that is already compiled for a specific native processor does not need to be interpreted and can save execution time by skipping this step.

Java Comments

- Single Line Comments – Single line comments start with two "//" slash characters.

  // This is a Single Line Java Comment

- Multi Line Comments – Multi line comments start with a "/*" slash followed by an asterisk and end in a "*/" an asterisk followed by a slash.

  ```
  /*
      This is
      a Multi-Line
      Comment
  */
  ```

Java Basic Data Types

- byte – A 8 bit number with values from –128 to 127 inclusive.

- short – A 16 bit number with values from -32,768 to 32,767 inclusive.

- int - A 32 bit number with values from -2,147,483,648 to 2,147,483,647 inclusive.

- long – A 64 bit number with values from -9,223,372,036,854,775,808 to 9,223,372,036,854,775,807 inclusive.

- float - A single-precision 32-bit IEEE 754 floating point number.

- double – A double-precision 64-bit IEEE 754 floating point number.

- char - A single 16-bit Unicode character that has a range of '\u0000' (or 0) to '\uffff' (or 65,535 inclusive).

- boolean – Has a value of either true or false.

Arrays

In Java you can create arrays of elements from the basic Java data types listed above. The following statement creates an array of m_DataSize elements of type byte.

int   m_DataSize = 700000;

byte[]   m_DataByte = new byte[m_DataSize];

Data Modifiers

- private – Variables that are private are only accessible from within the class they are declared in.
- public – Variables that are public can be accessed from any class.
- static – Variables that are declared static have only one copy associated with the class they are declared in.
- final – The final modifier indicates that the variable will not change.

Java Operators

Arithmetic Operators

- \+   Additive operator (also used for String concatenation)
- \-   Subtraction operator
- *   Multiplication operator
- /   Division operator
- %   Remainder operator

Unary Operators

- \+   Unary plus operator
- \-   Negates an expression
- ++   Increments a value by 1
- \-\-   Decrements a value by 1
- !   Inverts the value of a boolean

Conditional Operators

- &&   Conditional-AND
- || Conditional-OR
- =   Assignment operator
- ==   Equal to

- !=   Not equal to
- \>   Greater than
- \>=   Greater than or equal to
- <   Less than
- <=   Less than or equal to

Bitwise and Bit Shift Operators

- ~   Unary bitwise complement
- <<   Signed left shift
- \>>   Signed right shift
- \>>>   Unsigned right shift
- &   Bitwise AND
- ^   Bitwise exclusive OR
- | Bitwise inclusive OR

Java Flow Control Statements

- if then statement

    ```
    if (expression)
    {
            // execute statements here if expression evaluates to true
    }
    ```

- if then else statement

    ```
    if (expression)
    {
            // execute statements here if expression evaluates to true
    }
    else
    {
    ```

// execute statements here if expression evaluates to false

}

- **switch statement**

    switch(expression)

    {

        case label1:

            // Statements to execute if expression evaluates to

            // label1:

        break;

        case label2:

            // Statements to execute if expression evaluates to

            // label2:

        break;

    }

- **while statement**

    while (expression)

    {

        // Statements here execute as long as expression evaluates // to true;

    }

- **for statement**

    for (variable counter initialization; expression; variable counter increment/decrement)

    {

        // variable counter initialized when for loop is first

        // executed

        // Statements here execute as long as expression is true

// counter variable is updated

}

Java Classes

Java is an object oriented language. What this means is that you can derive or extend existing classes to form new customized classes of the existing classes. The derived class will have all the functionality of the parent class in addition to any new functions that you may want to add in.

The following class is a customized version of its parent class from which it derives which is the Activity class. The MainActivity class will be the main and most important class we will be dealing with on the Android side of this book. All of our other custom classes are executed directly or indirectly from the MainActivity class.

public class MainActivity extends Activity

{

   // Body of class

}

Packages and Classes

Packages are a way in Java to group together classes, and interfaces that are related in some way. For example, a package can represent a single application.

package com.example.bluetoothtest;

Accessing Classes in Packages

In order to access classes that are located in other packages you have to bring them into view using the "import" statement. For example, in order to use the Log class that is located inside the android.util.Log package you need to import it with the following statement.

import android.util.Log;

Then, you can use the class definition without the full package name such as:

Log.e("error", "This Language is not supported");

To print out an error message to the Log debug window in the Eclipse development environment.

Accessing Class Variables and Functions

You can access a class's variables and functions through the "." operator just like in C++. See some examples below.

BluetoothMessageHandler   m_BluetoothMessageHandler = null;

m_PhotoData = m_BluetoothMessageHandler.GetBinaryData();

m_BluetoothMessageHandler.ResetData();

m_BluetoothMessageHandler.SetCommand(Command);

String Data = m_BluetoothMessageHandler.GetStringData();

Java Functions

The general format for Java functions is the same as in other languages such as C/C++. The function heading starts with optional modifiers such as private, public, or static. Next is a return value that can be void if there is no return value or a basic data type or and class. This is followed by the function name and then the parameter list.

Modifiers Return_value FunctionName(ParameterType1 Parameter1, ...)

{

    // Code Body

}

Calling the Parent Function

A function in a derived class can override the function in the parent or superclass using the @Override annotation. This is not required but helps to prevent programming errors. If the intention is to override a parent function but the function does not in fact do this then a compiler error will be generated.

In order for the function in a derived class to actually call its corresponding function in the parent class you use the super prefix as seen below.

@Override

public void onCreate(Bundle savedInstanceState)

{

    super.onCreate(savedInstanceState);

    // Put new original code here

}

> Note: Additional Java tutorials can be found on http://docs.oracle.com/javase/tutorial/

Java Interfaces

The purpose of a Java interface is to provide a standard way for programmers to implement the actual functions in an interface in code in a derived class. An interface does not contain any actual code just the function definitions. The function bodies with the actual code must be defined by other classes that implement that interface. For example, the MainActivity class in the examples in this book implements the Runnable interface in order for code in other classes to change the Android's User Interface. In order to do this the run() function must be defined in the MainActivity class.

public class MainActivity extends Activity implements Runnable

{

    public void run()

    {

        // Original code to implement the Runnable interface

    }

}

Overview of Bluetooth for Android

Bluetooth is a wireless technology that is used for exchanging data over short distances. The Android operating system on both tablets and mobile phones support bluetooth.

Bluetooth Permissions

In order to use Bluetooth on Android you will need to declare certain permissions.

You will need to declare the BLUETOOTH permission in the "AndroidManifest.xml" file in order to perform any Bluetooth communication, such as requesting a connection, accepting a connection, and transferring data. To do this, add the following to the Android manifest file.

<uses-permission android:name=*"android.permission.BLUETOOTH"* />

You will need to declare the BLUETOOTH_ADMIN permission in the "AndroidManifest.xml" file in order to initiate device discovery or manipulate Bluetooth settings. To do this, add the following to the Android manifest file.

<uses-permission android:name=*"android.permission.BLUETOOTH_ADMIN"* />

Checking for Bluetooth Support

Before you start using Bluetooth on Android you should first check for the availability of Bluetooth.

Checking for Bluetooth support involves:

1. Checking to see if Bluetooth is supported on the Android device by calling BluetoothAdapter.*getDefaultAdapter*() and checking to see if the result is "null".

2. If the result is "null" then the device does not support Bluetooth.

3. If the result is not null then the device does support Bluetooth and a BlueToothAdapter object is returned which is assigned to m_BluetoothAdapter.

4. Next, if Bluetooth is supported you will need to check if Bluetooth is enabled by calling m_BluetoothAdapter.isEnabled().

5. If it is not enabled then enable it by creating a new Intent object and calling the startActivityForResult(enableBtIntent, REQUEST_ENABLE_BT) function with this new Intent object as a parameter and a user defined REQUEST_ENABLE_BT parameter. This generates a prompt for the user to choose if he wants to turn on Bluetooth. The REQUEST_ENABLE_BT is used in the onActivityResult() function to determine if the user clicked the "ok" button to turn on Bluetooth. If the user selected to turn on Bluetooth then the InitializeBlueTooth() function is called to initialize Bluetooth for Android.

6. If Bluetooth is already enabled on the device then call the InitializeBlueTooth() to initialize Bluetooth for this application.

See Listing 3-1.

Listing 3-1. Checking for Bluetooth Support

```
m_BluetoothAdapter = BluetoothAdapter.getDefaultAdapter();

if (m_BluetoothAdapter == null) {

    // Device does not support Bluetooth

    // Print error message

}
else
{

    // Device does support bluetooth

    if (!m_BluetoothAdapter.isEnabled())

    {

        // Bluetooth is not enabled so try to enable it.

        Intent enableBtIntent = new Intent(BluetoothAdapter.ACTION_REQUEST_ENABLE);

        startActivityForResult(enableBtIntent, REQUEST_ENABLE_BT);

    }
```

```java
            else
            {
                // Blue Tooth is Enabled so initialize it.
                InitializeBlueTooth();
            }
}


// Process enable bluetooth dialog
protected void onActivityResult (int requestCode, int resultCode, Intent data)
{
        if (requestCode == REQUEST_ENABLE_BT)
        {
                if (resultCode == RESULT_OK )
                {
                        m_DebugMsg += "User clicked on YES button to turn BlueTooth on!!!!!!!!!! \n";
                        m_DebugMsgView.setText(m_DebugMsg.toCharArray(), 0, m_DebugMsg.length());
                        Log.d("BlueToothTest","User Clicked YES to turn bluetooth On!!!!!!!!!");

                        // User wants to allow so initialize it.
                        InitializeBlueTooth();
                }
                else
                if (resultCode == RESULT_CANCELED)
                {
                        m_DebugMsg += "User clicked on NO button and has declined to start Bluetooth. Bluetooth NOT Started!!!!!!!!! \n";
                        m_DebugMsgView.setText(m_DebugMsg.toCharArray(), 0, m_DebugMsg.length());
                        Log.d("BlueToothTest","User declined to start Bluetooth. Bluetooth NOT Started!!!!!!!!!!");
                }
```

}

}

Initializing Bluetooth for Android

To initialize the Bluetooth on Android you need to:

1. Register a BroadcastReceiver object that will initialize the connection for the Bluetooth device found by calling RegisterReceiver().

2. Start the discovery for available Bluetooth devices to connect to by calling m_BluetoothAdapter.startDiscovery(). The startDiscovery() function returns true if discovery has started false if discovery has failed to start.

See Listing 3-2.

Listing 3-2. Initializing Bluetooth

```
void InitializeBlueTooth()
{
    // Keep track of Blue Tooth Active status
    m_BlueToothActive = true;

    // Register Broadcast Receiver
    RegisterReceiver();

    // Start the discovery of blue tooth enabled devices
    boolean success = m_BluetoothAdapter.startDiscovery();
    if (success)
    {
        Log.d("BlueToothTest", "BlueTooth Device Discovery Starting!!");
    }
    else
    {
        Log.d("BlueToothTest", "BlueTooth Device Discovery FAILED!!");
```

                }
        }

The RegisterReceiver() function creates and sets the actual BroadcastReceiver function that is called each time the startDiscovery() function finds a Bluetooth device.

The RegisterReciever function does the following:

1. Creates a new BroadcastReceiver function and sets it to the global m_Receiver variable and defines its onReceive() function.

2. The onReceive() function does the following:

    1. Gets the action type that caused the onReceive() function to be called by calling intent.getAction();

    2. Checks this action type to see if it indicates that a new Bluetooth device has been found and no previous Bluetooth devices have been found.

    3. If a new Bluetooth device has been found and this is the first one then get an object representing a Bluetooth device by calling intent.getParcelableExtra(BluetoothDevice.EXTRA_DEVICE) and setting it to the variable device.

    4. Gets the device name by calling device.getName().

    5. Get the device address by calling device.getAddress().

    6. Displays this information to the output message window on Android by calling the setText() function on the m_OutputMsg variable.

    7. Announces the device information by using the Android's built in text to speech capacity by calling the m_TTS.speak() function. The m_TSS is the global TextToSpeech `object that we will use for all speech in our examples`.

    8. Initializes a Bluetooth connection with the device that was just found by creating a new BlueToothClientConnectThread class object and setting it equal to the global variable m_ClientConnectThread.

    9. Starts the actual connection to the device by calling m_ClientConnectThread.start();

3. The newly created BroadCastReceiver object is actually activated by:

    1. Creating a new IntentFilter object by calling IntentFilter(BluetoothDevice.*ACTION_FOUND*) with the ACTION_FOUND parameter to indicate that this IntentFilter relates to the Bluetooth device being found. This new object is set to the local variable filter

    2. The registerReceiver(m_Receiver, filter) function is called with the parameters m_Receiver which holds the BroadCastReceiver object with the newly created IntentFilter filter.

See Listing 3-3.

Listing 3-3. The RegisterReciever() Function

```
void RegisterReceiver()
{
    // Create a BroadcastReceiver for ACTION_FOUND
    m_Receiver = new BroadcastReceiver() {
        public void onReceive(Context context, Intent intent) {
            String action = intent.getAction();
            // When discovery finds a device
            if (BluetoothDevice.ACTION_FOUND.equals(action) && !m_DeviceFound) {
                // Get the BluetoothDevice object from the Intent
                BluetoothDevice device = intent.getParcelableExtra(BluetoothDevice.EXTRA_DEVICE);

                // Output device name and address to window on Android
                m_OutputMsg += "";
                m_OutputMsg += device.getName();
                m_OutputMsg += "";
                m_OutputMsg += " ----- ";
                String DeviceName = device.getName();

                m_OutputMsg += "";
                m_OutputMsg += device.getAddress();
                m_OutputMsg += "";
                m_OutputMsg += "\n";

                m_OutputMsgView.setText(m_OutputMsg.toCharArray(), 0, m_OutputMsg.length());

                // The if statement can test here for a match in specfic device name and/or device address.
                // For simplicity we set it to true to connect to the first available bluetooth device
```

```java
                    if (true)
                    {
                        // Zombie Copter Controller Found!
                        String ZFound = "ZombieCopter Controller Found!!!\n";
                        m_TTS.speak(m_DeviceName + "Found", TextToSpeech.QUEUE_ADD, null);
                        m_TTS.speak("Device name is " + DeviceName, TextToSpeech.QUEUE_ADD, null);


                        AddDebugMessage(ZFound);
                        m_Device = device;
                        m_DeviceFound = true;


                        // Create Android Client/Master connection to BlueTooth Module
                        m_ClientConnectThread = new BlueToothClientConnectThread(m_Device,
                                    m_BluetoothAdapter,
                                    m_BluetoothMessageHandler,
                                    m_UUID,
                                    MainActivity.this);
                        m_ClientConnectThread.start();
                        m_DebugMsg += "Android BlueTooth Client Connection Thread Started!! \n\n";
                        m_DebugMsgView.setText(m_DebugMsg.toCharArray(), 0, m_DebugMsg.length());
                    }
                }
            }
    };
    // Register the BroadcastReceiver
    IntentFilter filter = new IntentFilter(BluetoothDevice.ACTION_FOUND);
    registerReceiver(m_Receiver, filter); // Don't forget to unregister during onDestroy
}
```

Connecting to another Device Using Bluetooth

The type of Bluetooth connection between the Android device and the Arduino device for the examples in this book is where the Android device serves as a Master/Client and the Arduino device serves as a Slave/Server.

A Master or Client device initiates the Bluetooth connection and the Slave or Server device listens for a request to connect via Bluetooth. The Android has a built in Bluetooth capability. The Arduino requires an additional Bluetooth adapter to be able to communicate using Bluetooth.

When the Android first discovers the Arduino's Bluetooth adapter it will try and "pair" with the device. What this means is that a popup screen will appear asking the user to type in a password which is usually "1234" that will allow the Android and Arduino to establish a Bluetooth connection.

We discuss the Arduino side of Bluetooth communication later in this chapter. See Figure 3-1.

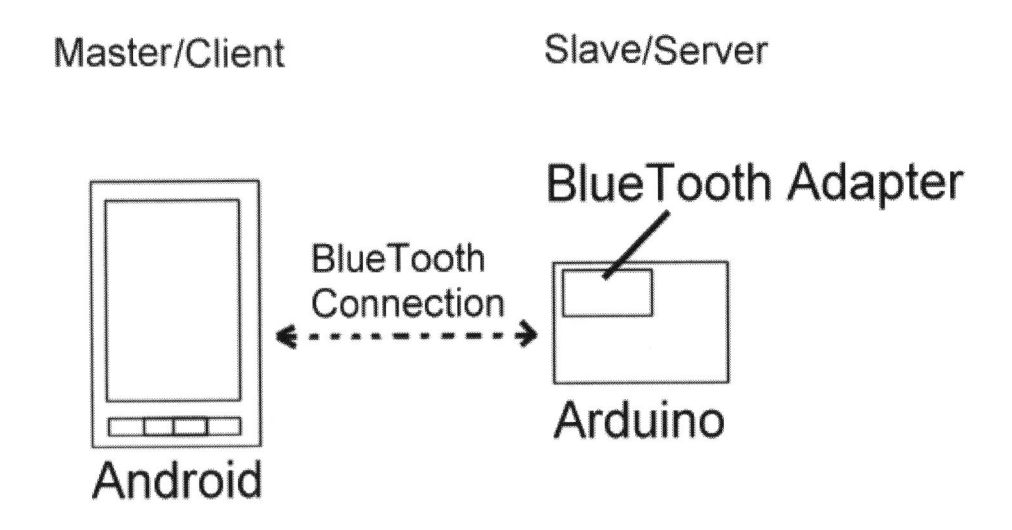

Figure 3-1. Our Bluetooth general connection setup

Custom Classes Used for Bluetooth Connection

This section covers the custom classes that are used to implement the actual bluetooth connection and to manage the connection between devices. The BlueToothClientConnectThread class is used to get a Bluetooth socket from the server or Arduino and then to connect to it. In order to connect to it and to manage the connection a new BluetoothServerConnectedThread class object is created and executed. Inside the BluetoothServerConnectedThread class is the BluetoothMessageHandler class which is the class that actually manages the connection between the Android and Arduino devices. See Figure 3-2.

```
┌─────────────────────────────────────┐
│      BlueToothClientConnectThread   │
└─────────────────────────────────────┘
                  │
                  ▼
┌─────────────────────────────────────┐
│    BluetoothServerConnectedThread   │
│   ┌─────────────────────────────┐   │
│   │   BluetoothMessageHandler   │   │
│   └─────────────────────────────┘   │
└─────────────────────────────────────┘
```

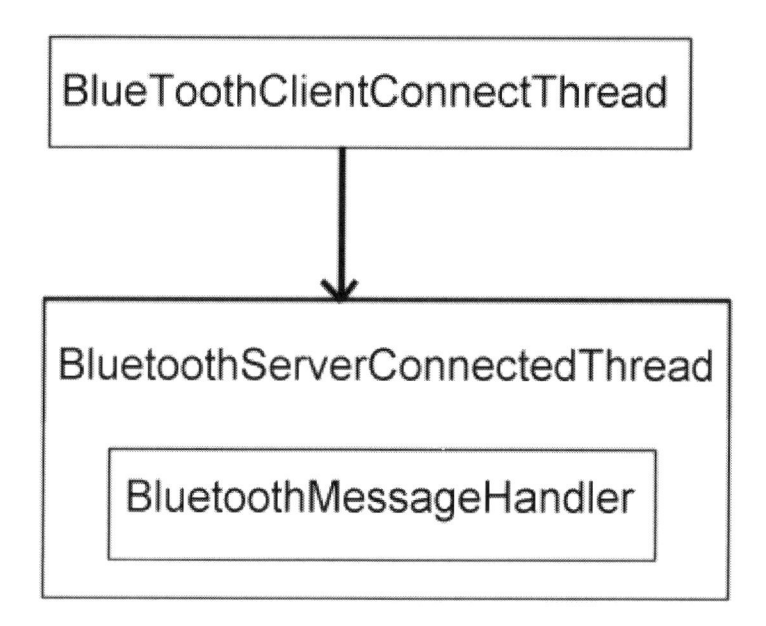

Figure 3-2. Our custom Bluetooth connection related classes

The BlueToothClientConnectThread Class

The BlueToothClientConnectThread class is the main entry point for creating a Bluetooth connection to another device. The class derives or extends from the Thread class which is a standard built in Android class.

public class BlueToothClientConnectThread extends Thread

A new BlueToothClientConnectThread class object is created in the RegisterReceiver() function discussed previously and is set to the global variable m_ClientConnectThread which is the main access point in sending commands to the Arduino using a Bluetooth connection.

In creating the BlueToothClientConnectThread class object the uuid parameter is defined as the following which is the standard UUID for a Bluetooth Serial board module that is part of the SPP or "Serial Port Profile".

UUID m_UUID = UUID.fromString("00001101-0000-1000-8000-00805F9B34FB");

The class member data is as follows:

private final BluetoothSocket m_Socket;

private final BluetoothDevice m_Device;

BluetoothAdapter m_BluetoothAdapter = null;

BluetoothServerConnectedThread m_ConnectedThread = null;

BluetoothMessageHandler m_BluetoothMessageHandler = null;

MainActivity m_MainActivity = null;

The constructor of this class which is

public BlueToothClientConnectThread(BluetoothDevice device,

        BluetoothAdapter iBlueToothAdapter,

        BluetoothMessageHandler iMessageHandler,

        UUID uuid,

        MainActivity iActivity)

is executed when this object is created. The device parameter is the newly discovered Bluetooth device. The iBlueToothAdapter parameter is the Android's Bluetooth communication module that will communicate with the device. The iMessageHandler parameter is the BluetoothMessageHandler object that processes the return data from the commands that are sent to the Arduino. The uuid parameter is the Serial Port Profile UUID value mentioned previously. The iActivity parameter is an object that references the current MainActivity object.

The constructor does the following:

1. Initializes key global class variables with the constructor's input parameters so they can be used throughout the class.

2. A socket is requested from the Arduino by calling the device.createRfcommSocketToServiceRecord(uuid) function with the SPP or Serial Port Profiles UUID.

3. Once the socket is received it is held in the m_Socket global class variable.

See Listing 3-4.

Listing 3-4. The BlueToothClientConnectThread Constructor

```
public BlueToothClientConnectThread(BluetoothDevice device,

                BluetoothAdapter iBlueToothAdapter,

                BluetoothMessageHandler iMessageHandler,

                UUID uuid,

                MainActivity iActivity) {


        m_BluetoothAdapter = iBlueToothAdapter;

        m_BluetoothMessageHandler = iMessageHandler;

        m_MainActivity = iActivity;



        // Use a temporary object that is later assigned to m_Socket,
```

```
            // because m_Socket is final

            BluetoothSocket tmp = null;

            m_Device = device;


            // Get a BluetoothSocket to connect with the given BluetoothDevice

            try {

                    // MY_UUID is the app's UUID string, also used by the server code

                    tmp = device.createRfcommSocketToServiceRecord(uuid);

            } catch (IOException e) { }

            m_Socket = tmp;
}
```

The run() function connects the Android to the socket received from the Arduino in the constructor and initiates the management of the connection.

Specifically the run() functions does the following:

1. Cancels the discovery process to find new Bluetooth enabled devices by calling m_BluetoothAdapter.cancelDiscovery().

2. Connects the Android to the Arduino using Bluetooth by calling m_Socket.connect(). If there is an error in the connection process then the socket is closed.

3. The connection is then managed by calling the ManageConnectedSocket(m_Socket) function with the m_Socket variable as a parameter

See Listing 3-5.

Listing 3-5. The run() function

```
public void run() {

            // Cancel discovery because it will slow down the connection

            m_BluetoothAdapter.cancelDiscovery();


            try {

                // Connect the device through the socket. This will block

                // until it succeeds or throws an exception
```

```
                    m_Socket.connect();

            } catch (IOException connectException) {

                // Unable to connect; close the socket and get out

                try {

                    m_Socket.close();

                } catch (IOException closeException) { }

                return;

            }

            // Do work to manage the connection (in a separate thread)

            ManageConnectedSocket(m_Socket);

}
```

The ManageConnectedSocket() function does the following:

1. Creates a new BluetoothServerConnectedThread object with the connected Socket and the m_BluetoothMessageHandler as parameters and assigns this to the m_ConnectedThread global class variable.

2. The incoming data from the Arduino is read in and processed by calling m_ConnectedThread.start().

3. The MainActivity class is notified that Bluetooth is now active by calling m_MainActivity.BlueToothConnected().

4. The MainActivity class is notified to activate the "Take Photo" button on the Android user interface by calling m_MainActivity.SetTakePhotoButtonActive().

5. The Android's user interface is actually updated by calling m_MainActivity.runOnUiThread(m_MainActivity) which executes the MainActivity's run() function in order to do the actual update of the user interface.

See Listing 3-6.

Listing 3-6. The ManageConnectedSocket() function

```
void ManageConnectedSocket(BluetoothSocket Socket)

{

        Log.e("BlueToothClientConnectThread", "In ManageConnectedSocket");


        m_ConnectedThread = new BluetoothServerConnectedThread(Socket, m_BluetoothMessageHandler);
```

```
            m_ConnectedThread.start();


            m_MainActivity.BlueToothConnected();

            m_MainActivity.SetTakePhotoButtonActive();


            // Update the User Interface

            m_MainActivity.runOnUiThread(m_MainActivity);
}
```

The GetConnectedThread() function returns the m_ConnectedThread variable that is a BluetoothServerConnectedThread object. With this object you can write commands to the Arduino using Bluetooth. This thread also handles the reading of data from the Arduino.

See Listing 3-7.

Listing 3-7. The GetConnected() function

```
BluetoothServerConnectedThread GetConnectedThread()
{
        return m_ConnectedThread;
}
```

The Cancel() function will cancel the current connection and close the Bluetooth connection by calling the m_Socket.close() function.

See Listing 3-8.

Listing 3-8. The cancel() function

```
/** Will cancel an in-progress connection, and close the socket */
public void cancel() {
        try {
            m_Socket.close();
        } catch (IOException e) { }
}
```

The BluetoothServerConnectedThread Class

The BluetoothServerConnectedThread class is where data is read in from the Arduino and written to the Arduino using Bluetooth. The class is derived from or extends from the Thread class which is a standard built in Android class.

public class BluetoothServerConnectedThread extends Thread

The class member data is as follows:

private final BluetoothSocket m_Socket;

private final InputStream   m_InStream;

private final OutputStream   m_OutStream;

BluetoothMessageHandler m_BluetoothMessageHandler;

The class constructor is called with parameters socket which is the Bluetooth connection socket and BluetoothMessageHandler which is the class object that will handle the data received back from the Arduino as a result of commands sent by the Android controller.

BluetoothServerConnectedThread(BluetoothSocket socket,

        BluetoothMessageHandler BluetoothMessageHandler)

The constructor does the following:

1. Initializes global class variables with the input parameters.

2. Declares and initializes some needed temporary variables.

3. The input Bluetooth data stream is set by calling the socket.getInputStream() function and the returned object is assigned to the global m_InStream variable.

4. The output Bluetooth data stream is set by calling the socket.getOutputStream() function and the returned object is set to the global m_OutStream.

See Listing 3-9.

Listing 3-9. The BluetoothServerConnectedThread constructor

public BluetoothServerConnectedThread(BluetoothSocket socket,

        BluetoothMessageHandler BluetoothMessageHandler)

{

    m_Socket = socket;

    InputStream tmpIn = null;

    OutputStream tmpOut = null;

    m_BluetoothMessageHandler = BluetoothMessageHandler;

```
// Get the input and output streams, using temp objects because
// member streams are final
try {
    tmpIn = socket.getInputStream();
    tmpOut = socket.getOutputStream();
} catch (IOException e) { }

m_InStream = tmpIn;
m_OutStream = tmpOut;
}
```

The run() function reads in the data from the Arduino and calls the bluetooth handler to process it.

The run() function does the following:

1. It creates an array of 1024 bytes that serves as a buffer for data being read into the Android from the Arduino.

2. Creates a variable called bytes that serves to hold the number of valid bytes in the byte buffer.

3. Reads continually from the input Bluetooth connection by calling m_InStream.read(buffer).

4. Handles this incoming data in the buffer by calling the m_BluetoothMessageHandler.ReceiveMessage(bytes, buffer) function with the number of bytes received and the buffer.

See Listing 3-10.

Listing 3-10. The run() function

```
public void run() {
    byte[] buffer = new byte[1024];  // buffer store for the stream
    int bytes; // bytes returned from read()

    // Keep listening to the InputStream until an exception occurs
    while (true) {
```

```
            try {

                // Read from the InputStream

                bytes = m_InStream.read(buffer);

                m_BluetoothMessageHandler.ReceiveMessage(bytes, buffer);

            } catch (IOException e)

            {

                break;

            }

        }

}
```

The function write() writes out data to the Arduino from the Android by:

1. Calling the m_OutStream.write(bytes) function with the array of bytes that were input as a parameter to the function. This is how we send commands to the Arduino from the Android controller such as "TakePhoto", "VGA", "QVGA", etc.

See Listing 3-11.

Listing 3-11. The write() function

```
/* Call this from the main activity to send data to the remote device */
public void write(byte[] bytes) {

            try {

                m_OutStream.write(bytes);

            } catch (IOException e)

            {

            }

}
```

The cancel() function closes the Bluetooth connection by:

1. Calling the m_Socket.close() function.

See Listing 3-12.

Listing 3-12. The cancel() function

```
/* Call this from the main activity to shutdown the connection */
public void cancel() {
    try {
        m_Socket.close();
    } catch (IOException e) { }
}
```

The BluetoothMessageHandler Class

The BluetoothMessageHandler class processes the data returned from the Arduino based upon the last command issued by the Android controller. The incoming data is in either text format or binary format.

Text Data

The text data format consists of numbers and letters followed by a newline character which is "\n" that indicates the end of the text data. See Figure 3-3.

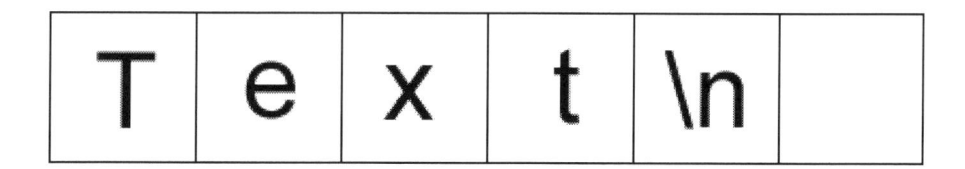

| T | e | x | t | \n | |

Figure 3-3. Text data

We specify the exact character to indicate the end of text using the variable m_EndData.

```
char    m_EndData    = '\n';
```

The text data is read in by calling the ReceiveTextData() function

```
boolean ReceiveTextData(int NumberBytes, byte[] Message)
```

Binary Data

The binary data is a series of 1's and 0's and the length of the returned data must be known when issuing an Android command that expects binary data returned from the Arduino. See Figure 3-4.

```
┌───┬───┬───┬───┬───┬───┐
│ 0 │ 1 │ 1 │ 0 │ 0 │ 1 │
└───┴───┴───┴───┴───┴───┘
```

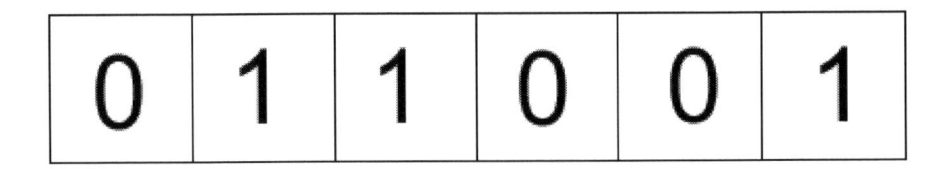

Figure 3-4. Binary Data

The data length in bytes of the incoming binary data is held in the m_DataIncomingLength variable that is declared as:

int   m_DataIncomingLength = 0;

The binary data is read in using the ReceiveBinaryData() function that is declared as follows:

boolean ReceiveBinaryData(int NumberBytes, byte[] Message)

Class Overview

The ReceiveMessage() function is the main entry point to this class and is called from the BluetoothServerConnectedThread class. This section gives you a general overview of how this class works and how you can add your own customizations to this class.

The general procedure to handle data that is being sent from the Arduino in response to an Android command is:

1. The BluetoothServerConnectedThread class object calls the ReceiveMessage() function with the data received from the Arduino and the number of bytes that data consists of.

2. The ReceiveMessage() function processes the data based on the last Android command issued to the Arduino and calls a function ProcessXXXXCommand(NumberBytes, Message) where XXXX is replaced by the command name.

3. If the data to be received is text data then the ReceiveTextData() function is used to process the data.

4. If the data to be received is binary data then the ReceiveBinaryData() function is used to process the data.

5. Once all the text or binary data is received then you need to set a variable in the MainActivity class to indicate that the Android command has received a response from the Arduino. You do this through a command such as m_MainActivity.SetXXXXFinished() where the XXXX is replaced by the command that has just received a response from Arduino.

6. In order to change the part of the Android user interface that the command updates you need to call m_MainActivity.runOnUiThread(m_MainActivity). This executes the run() function in the MainActivity class where the MainActivity class's user interface objects can be accessed.

See Figure 3-5.

```
                ┌─────────────────────────────────────────────┐
                │ ReceiveMessage(int NumberBytes, byte[] Message) │
                └─────────────────────────────────────────────┘
                                    │
                                    ▼
                ┌─────────────────────────────────────────────┐
                │ ProcessXXXXCommand(NumberBytes, Message)    │
                └─────────────────────────────────────────────┘
                                    │
                  Yes               ▼              No
              ┌──────────  Is Data Recieved Binary?  ──────────┐
              ▼                                                 ▼
  ┌──────────────────────────────────┐   ┌────────────────────────────────────────────┐
  │ ReceiveBinaryData(NumberBytes, Message) │   │ ReceiveTextData(int NumberBytes, byte[] Message) │
  └──────────────────────────────────┘   └────────────────────────────────────────────┘
                     │                                          │
                     └──────────────────┬───────────────────────┘
                                        ▼
                        ┌──────────────────────────────────┐
                        │ m_MainActivity.SetXXXXFinished(); │
                        └──────────────────────────────────┘
                                        │
                                        ▼
                  ┌────────────────────────────────────────────┐
                  │ m_MainActivity.runOnUiThread(m_MainActivity); │
                  └────────────────────────────────────────────┘
```

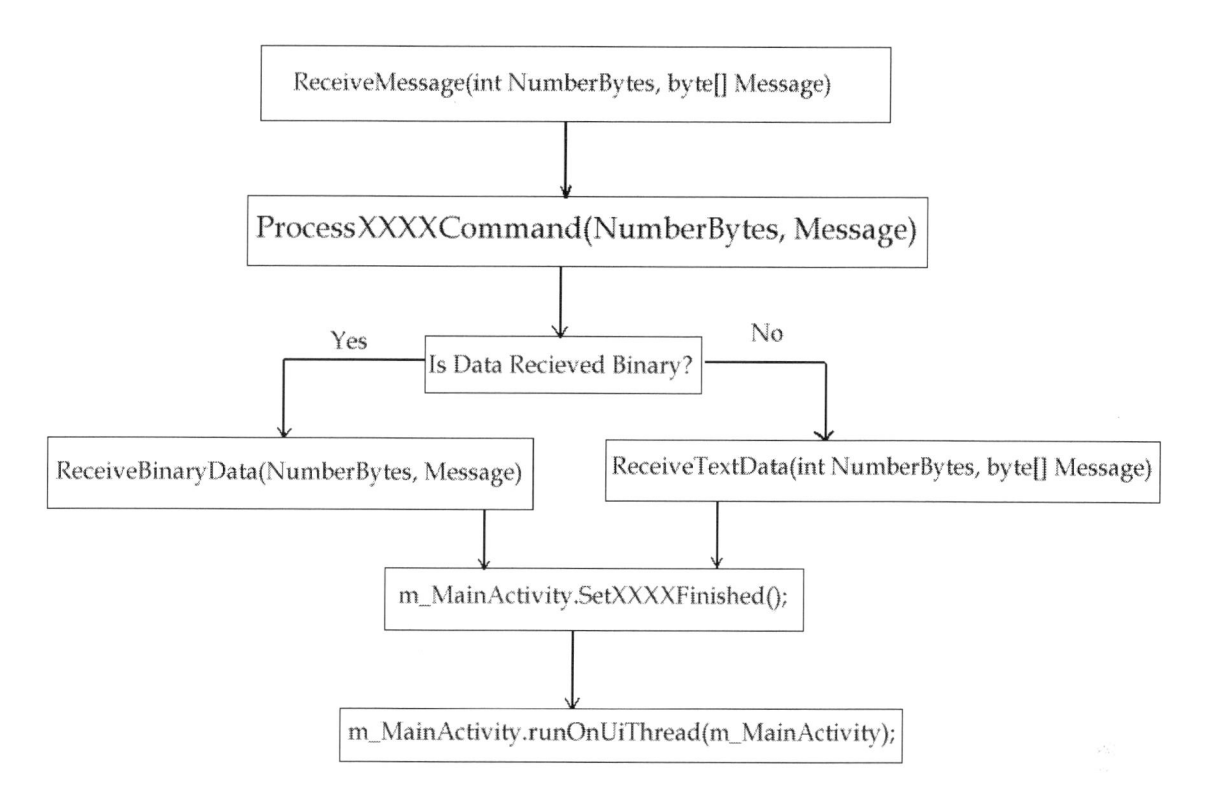

Figure 3-5. BluetoothMessageHandler Class flowchart

The m_Command is a String variable that holds an alphanumeric representation of the Android command that the Android expects to receive a response to. The m_Command variable is set using the SetCommand() function. See Listing 3-13.

Listing 3-13. The SetCommand() function

String m_Command = "";

void SetCommand(String Command)

{

    m_Command = Command;

}

The m_MainActivity variable holds a reference to the MainActivity class object that created this BluetoothMessageHandler object.

MainActivity m_MainActivity;

The m_USAscii variable holds the character set to use when converting incoming bytes into alphanumeric text.

Charset   m_USAscii   = null;

The m_Data variable holds text data

String   m_Data   = "";

The m_DataIncomingLength variable holds the length in bytes of the data expected to be sent by the Arduino.

int   m_DataIncomingLength = 0;

The m_DataByte array is an array of bytes of size m_DataSize that holds the binary data that will be sent by the Arduino.

byte[]   m_DataByte = new byte[m_DataSize];

The m_DataSize variable holds the size of the binary data buffer used to store incoming binary data and is set to 700,000 bytes long.

int   m_DataSize = 700000;

The m_DataIndex variable holds the index into the m_DataByte array and indicates the next available empty position in the array that can hold a new value.

int   m_DataIndex = 0;

The SetDataReceiveLength() function is called from the MainActivity class and sets the length of the expected incoming binary data in bytes.

void SetDataReceiveLength(int length) { m_DataIncomingLength = length;}

The GetBinaryData() function returns a reference to m_DataByte which holds binary data sent from the Arduino.

byte[] GetBinaryData(){return m_DataByte;}

The GetBinaryDataLength() function returns the length in bytes of the data in the m_DataByte array.

int GetBinaryDataLength() { return m_DataIndex;}

The GetStringData() function returns the text data sent from the Arduino that is stored in m_Data. The m_Data variable is also reset to the null string. See Listing 3-14.

Listing 3-14. GetStringData() function

// Get String Data and resets internal String Data variable

String GetStringData()

{

   String temp = m_Data;

   m_Data = "";

return temp;

}

The ResetData() function initializes the text and binary data structures that are used to receive data from the Arduino by:

1. Setting m_Data which holds the incoming text data to "".

2. Setting m_DataIndex that holds the position of the next available byte in the incoming binary data buffer to 0 that is the beginning of the buffer.

3. Setting m_DataIncomingLength that holds the length in bytes of the expected binary data from Arduino to 0.

4. Erasing the existing data in the binary data buffer m_DataByte by writing 0 to every array location.

See Listing 3-15.

Listing 3-15. The ResetData() function

```
void ResetData()
{
    // Reset String Data
    m_Data = "";

    // Reset Binary Data
    m_DataIndex = 0;
    m_DataIncomingLength = 0;

    // Erase old data in array
    for (int i = 0 ; i < m_DataSize; i++)
    {
        m_DataByte[i] = 0;
    }
}
```

The class constructor initializes the class object by:

1. Assigning the global variable m_MainActivity to the Actvity class object that will use this handler.

2. Prints out a message to the Android debug window indicating that the constructor has been executed.

3. Assigns the US ASCII character set to m_USAscii by calling Charset.forName("US-ASCII"). The m_USAscii variable is used to convert incoming text data into US English characters.

See Listing 3-16.

Listing 3-16. The Constructor

```
BluetoothMessageHandler(MainActivity iActivity)
{
    m_MainActivity = iActivity;

    m_Message = "BlueToothMessageHandler Initialized!!\n";

    m_MainActivity.AddDebugMessage(m_Message);


    // Set Charset to US ASCII translation
    m_USAscii = Charset.forName("US-ASCII");
}
```

The ReceiveTextData() function reads in the data from the Message byte array input parameter and returns true if a complete text data message has been received and false otherwise.

Specifically, the function does the following:

1. Read in the valid data from the Message array based on the NumberBytes input parameter.

2. Convert this data into a ByteBuffer object using the ByteBuffer.wrap(temp) function with temp being the valid data from Step 1.

3. Convert the ByteBuffer object into a CharBuffer object by calling the m_USAscii.decode(bb) function with the input parameter bb being the ByteBuffer obtained in Step 2.

4. Convert the CharBuffer object obtained in the previous step into a String object by calling the TempCharBuffer.toString() function on the CharBuffer object.

5. Add the String obtained from Step 4 to the current text data message which is stored in m_Data.

6. Next, we check to see if the end of the text message has been reached. We first find the array index value of the text end message character that is stored in m_EndData by calling the m_Data.indexOf(m_EndData) function.

7. If the returned value is greater than or equal to 0 then we have found the end of text message marker and this text message is complete.

See Listing 3-17.

Listing 3- 17. The ReceiveTextData() function

```
// Receives Text Data from Bluetooth
boolean ReceiveTextData(int NumberBytes, byte[] Message)
{
    boolean EndTextDataFound = false;
    byte[] temp = new byte[NumberBytes];


    // Capture readable data
    for (int i = 0; i < NumberBytes; i++)
    {
        temp[i] = Message[i];
    }
    ByteBuffer bb = ByteBuffer.wrap(temp);


    // Convert from bytes to printable characters
    CharBuffer TempCharBuffer = m_USAscii.decode(bb);


    // Debug output Number of bytes incoming and characters
    m_Message = TempCharBuffer.toString();
    Log.d("ZombieCopter Text Recieve NumberBytes " , NumberBytes + "");
    Log.d("ZombieCopter Text Recieve Partial Message",  m_Message);


    // Add Partial Message to Complete Data
    m_Data += m_Message;
```

```
        // Check if end of data has been reached
        int EndOfData = m_Data.indexOf(m_EndData);
        if (EndOfData >= 0)
        {
            // All Text Data Has been Read in
            EndTextDataFound = true;
        }
        return EndTextDataFound;
}
```

The ReceiveBinaryData() function reads in the incoming binary data from the Arduino.

The function does the following:

1. Reads in the incoming binary data from the Message byte array and stores it in the m_DataByte array. The m_DataIndex variable that indicates the next empty position in the m_DataByte array, is incremented.

2. If m_DataIndex is greater than or equal to the expected length of the incoming binary message which is held in m_DataIncomingLength then the incoming binary message has completed. Otherwise, more data needs to be read in from the Arduino.

See Listing 3-18.

Listing 3-18. The ReceiveBinaryData() function

```
boolean ReceiveBinaryData(int NumberBytes, byte[] Message)
{
        boolean Finished = false;

        // Add incoming Binary data to bytes data array
        for (int i = 0; i < NumberBytes; i++)
        {
                m_DataByte[m_DataIndex] = Message[i];
                m_DataIndex++;
```

```
            }


            // Check to see if all binary data has been received

            if (m_DataIndex >= m_DataIncomingLength)

            {

                    Finished = true;

            }

            return Finished;

}
```

The ProcessTakePictureCommand() function processes the data received from the Arduino after the Android issues a "TakePhoto" command. This function is a good example of how to deal with Android commands that request binary data from the Arduino. In this case it is an image from the camera that is attached to the Arduino.

The function does the following:

1. Receives the binary data by calling the ReceiveBinaryData(NumberBytes, Message) function with parameters Message which is the array of bytes that holds the incoming message and NumberBytes which holds the number of valid bytes in the Message array.

2. If the ReceiveBinaryData() function returns true then the binary data message transfer has completed.

3. If the binary data message transfer has completed then:

    1. Notify the MainActivity class that the "TakePhoto" command has a completed response from the Arduino by calling the m_MainActivity.TakePhotoCommandCallback() function.

    2. The run() function in the MainActivity class is executed by calling the m_MainActivity.runOnUiThread(m_MainActivity) function. This updates the Android user interface so that the binary data is converted to a picture and displayed on the Android screen.

See Listing 3-19.

Listing 3-19. The ProcessTakePictureCommand() function

```
void ProcessTakePictureCommand(int NumberBytes, byte[] Message)

    {

            boolean FinishedReceivingData = ReceiveBinaryData(NumberBytes, Message);

            if (FinishedReceivingData)
```

{

// Process Take Picture Command

m_MainActivity.TakePhotoCommandCallback();

Log.d("BlueTooth Test","PhotoDataReadyCallback() called. Photo Loaded in ... ");

// Update the User Interface

m_MainActivity.runOnUiThread(m_MainActivity);

}

}

The ProcessDirCommand() function processes the Android command to get the directory of files on the Arduino SD Card. This function is a good example of how to handle an Android command that expects to receive text data from the Arduino.

The function does the following:

1. Receives incoming text data by calling the ReceiveTextData(NumberBytes, Message) function with Message that is the byte array buffer that holds the data and NumberBytes that holds the length of the data within the buffer.

2. If the ReceiveTextData() function returns true then the text data has completed its transmission. Otherwise the result is false.

3. If the text data transfer has completed then:

    1. The MainActivity class is notified that the incoming text data is ready to be processed by calling the m_MainActivity.SetDirectoryReadFinished() function.

    2. The Android user interface is updated and the new filenames are loaded into the appropriate menu by executing the run() function in the MainActivity class by calling the m_MainActivity.runOnUiThread(m_MainActivity) function.

See Listing 3-20.

Listing 3-20. The ProcessDirCommand() function

```
void ProcessDirCommand(int NumberBytes, byte[] Message)

{

    boolean FinishedReceivingText = ReceiveTextData(NumberBytes, Message);

    if (FinishedReceivingText)

    {
```

```
            // Process Remote Directory Command

            m_MainActivity.SetDirectoryReadFinished();

            Log.d("BlueTooth Test","SetDirectoryReadFinished() called. Arduino SD Card Directory Loaded in ... ");


            // Update the User Interface

            m_MainActivity.runOnUiThread(m_MainActivity);

      }
}
```

The ReceiveMessage() function is called whenever data from the Arduino is received by the Android using Bluetooth.

The function does the following:

1. If the command that was sent to the Arduino was a "TakePhoto" command then the ProcessTakePictureCommand() function is called.

2. If the command that was sent to the Arduino was the list directory command or "dir" then the ProcessDirCommand() function is called.

3. If the string held in m_Command does not match either the "TakePhoto" command or the "dir" command then an error is printed in the Android developer log console by calling the Log.e() function.

See Listing 3-21.

Listing 3-21. The ReceiveMessage() Function

```
void ReceiveMessage(int NumberBytes, byte[] Message)

{

      // Process Incoming Data

      // Assume incoming data is associated with the current m_Command variable

      // and process the Message accordingly.

      if (m_Command == "TakePhoto")

      {

            ProcessTakePictureCommand(NumberBytes, Message);

      } else

      if (m_Command == "dir")
```

```
            {
                    ProcessDirCommand(NumberBytes, Message);
            }
            else
            {
                    Log.e("BlueToothTest" , "Error - Command for Data Receive Not Found!!!!!!");
            }
}
```

Cleaning Up Bluetooth on Application Exit

In terms of exiting the Android application you need to add certain code to the onDestroy() function in the MainActivity class.

The modified onDestroy() function does the following:

1. Calls the parent onDestroy() function by calling super.onDestroy().

2. If Bluetooth is active then do the following:

   1. Unregister the broadcast receiver function that was registered during the initialization phase by calling the unregisterReceiver(m_Receiver) function with the broadcast receiver object held in m_Receiver.

   2. If the ClientConnectThread object has been initialized and is therefore not equal to null then:

      1. If there is an active connection between the Arduino and Android (m_ClientConnectThread.GetConnectedThread() != null) then call the cancel() function on that thread object to close the Bluetooth socket.

      2. Even if there is not an active connection close the Bluetooth socket that you opened previously in initializing Bluetooth by calling the m_ClientConnectThread.cancel() function.

See Listing 3-22.

Listing 3-22. The onDestroy() function for Bluetooth

```
@Override
protected void onDestroy()
{
        Log.d("BlueToothTest", "In onDestroy() FUNCTION !!!!!!!!");
```

```
            super.onDestroy();


    if (m_BlueToothActive)

    {

        Log.d("BlueToothTest", "Unregistering Broadcast Receiver!!!!!!!!");

        unregisterReceiver(m_Receiver);


        if (m_ClientConnectThread != null)

        {

            BluetoothServerConnectedThread ConnectedThread = m_ClientConnectThread.GetConnectedThread();

            if (ConnectedThread != null)

            {

                // Close Server Connected Thread

                Log.d("BlueToothTest", "Cancelling Connected Thread!!!!!!!!");

                ConnectedThread.cancel();

            }


            Log.d("BlueToothTest", "Cancelling Client Connect Thread!!!!!!!!");

            m_ClientConnectThread.cancel();

        }

    }

}
```

Overview of Bluetooth for Arduino

This section discusses the Bluetooth Adapter module that is used with the Arduino in order to connect with the Android device via Bluetooth. This section discusses the physical characteristics of the module as well as how to initialize it from the Arduino.

The Bluetooth Adapter

The Bluetooth adapter is connected to the Arduino and is the device that actually communications with the Android's Bluetooth.

The Bluetooth Adapter I use for this book was bought on Amazon.com. Specifically the item is described as:

KEDSUM® Arduino Wireless Bluetooth Transceiver Module Slave 4Pin Serial + DuPont Cable

The default communication rate of this module is 9600 baud.

> Note: The Arduino code for the projects in this book assumes a baud rate of 9600. If you are using a different baud rate then you will need to change the speed setting in the Arduino code.

The front of the Bluetooth adapter is shown in Figure 3-6.

Figure 3-6. Front of Bluetooth Adapter Module for Arduino

The back of the adapter is shown in Figure 3-7.

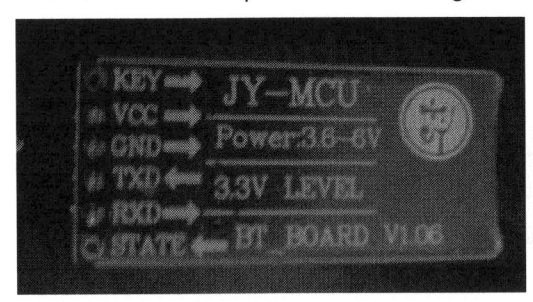

Figure 3-7. Back of Bluetooth Adapter Module for Arduino

The pins on the adapter are:

- VCC – Connects to the 3.3 V power output on the Arduino to power the device.
- GND – Connects to on of the GND pins on the Arduino.
- TXD – The pin that transmits data and that connects to the RXD pin on the Arduino.
- RXD – The pin that receives data and that connects to the TXD pin on the Arduino.

Initializing Bluetooth on Arduino

Bluetooth on the Arduino side is executed as Serial communication between Arduino and the Bluetooth adapter module that then communicates with the Bluetooth adapter on the Android device.

For our examples we will use the Software Serial library which allows for Serial communication between the Arduino and a device such our Bluetooth adapter module. The Software Serial library allows you to use pins for serial communication that are not specifically designated as serial communication pins (labeled rx and tx). In order to use this library we must include it in our Arduino program.

#include <SoftwareSerial.h>

In order initialize software serial we need to specify which Arduino pin is to serve as the Rx or receive data pin and which pin is to serve as the Tx or transmit data pin. Below we specify the Rx pin as 6 and the Tx pin as 7.

#define RxD 6

#define TxD 7

In order to initialize a new software serial class object we need to declare it with the Rx and Tx pins as input parameters to the class's constructor

SoftwareSerial BT(RxD,TxD);

The InitializeBlueTooth() function sets up the Bluetooth connection on the Arduino for use by:

1. Starting the software serial connection and setting the speed to 9600 baud by calling the BT.begin(9600) function.

2. Waiting for the serial connection to initialize by calling the delay(1000) which suspends execution of the Arduino code for 1 second or 1000 milliseconds.

3. Initialization message printed to Serial Monitor by calling the Serial.println() function.

4. Further initialization done by calling the initialize_bluetooth() function.

5. Waiting for initialization to complete by calling the delay(1000) function.

See Listing 3-22.

Listing 3-22. The InitializeBlueTooth() function

void InitializeBlueTooth()

{

// Initialize Bluetooth

BT.begin(9600);

```
  delay(1000);

  Serial.println(F("Initializing Bluetooth ..."));

  initialize_bluetooth();

  delay(1000);

}
```

The initialize_bluetooth() function checks to see if the Bluetooth adapter is working correctly by:

1. Calling the SendATCommand() function.

See Listing 3-23.

Listing 3-23. Initializing Bluetooth

```
void initialize_bluetooth()

{

    SendATCommand();

    // Other initializations such as set module name etc.

}
```

The SendATCommand() function checks to see if the Bluetooth adapter module is working correctly by sending an AT command to the module via the print() function. The returned characters should be "OK" if the bluetooth module is working correctly.

Specifically the function does the following:

1. Sends an AT command to the Bluetooth adapter by calling the BT.print(F("AT")) function which sends the character string "AT" to the adapter using our software serial class object BT.

2. Makes sure the data has been transmitted before continuing execution of the program by calling the BT.flush() function.

3. Delay execution of the program by 1500 milliseconds by calling the delay(1500) function

4. Test to see if the command was successful by calling the print_bt_response() function. If the function returns 1 then the command was successful and the Bluetooth adapter is working. If the function returns 0 then the Bluetooth adapter does not work correctly.

See Listing 3-24.

Listing 3-24.

```
void SendATCommand()

{

  int flag = 0;


  // AT Test Command

  BT.print(F("AT"));

  BT.flush();

  delay(1500);


  flag = print_bt_response();

  if(flag == 1)

  {

    Serial.print(F("AT Command Test ... Success\n\n"));

  }

  else if(flag == 0)

  {

    Serial.print(F("AT Command Failed\n\n"));

  }

}
```

The print_bt_response() function prints out the response from the Bluetooth adapter to the Serial Monitor and returns a value indicating if the previous command to the adapter worked correctly.

The function does the following:

1. While there is incoming data from the Bluetooth adapter, that is as long as BT.available()>0 evaluates to true do the following:

    1. Read in a single character from the incoming Bluetooth data stream by calling the (char)BT.read() function.

    2. Write out the character to the Serial Monitor by calling the Serial.print(out) function with the out input parameter being the character read in the previous step.

    3. If the previous character that was read in was a 'O' and the current character that was read in is a "K" then return 1.

    4. Otherwise, assign the current character to the outprev variable that holds the value of the character that was previously read in.

5. Continue reading from the Bluetooth serial data stream.

2. Return a 0 value.

See Listing 3-25.

Listing 3-25. The print_bt_response() function

```
int print_bt_response()
{
   int response;
   char out,outprev = '$';
   while(BT.available()>0)
   {
      out = (char)BT.read();
      Serial.print(out);
      if ((outprev == 'O')&&(out == 'K'))
      {
         Serial.print("\n");
         response = 1;
         return response;
      }
      outprev = out;
   }
   response = 0;
   return response;
}
```

The SetBlueToothModuleName() function is an optional function that the user can use to set the name of the Bluetooth module.

The function does the following:

1. Changes the name of the Bluetooth adapter module by calling the BT.print(F("AT+NAMEZombieCopter")) function with the parameter that include the new name to be set. In this case the new name is "ZombieCopter".

2. Makes sure all the data is transmitted to the Bluetooth adapter before continuing by calling the BT.flush() function.

3. Waits for 1500 milliseconds by calling the delay(1500) function.

4. Checks if the module name change was successful by calling the print_bt_response() function. If the return value is 1 then the name change was successful. If the return value is 0 then the name change was not successful.

See Listing 3-26.

Listing 3-26. The SetBlueToothModuleName() function

```
void SetBlueToothModuleName()
{
    // Setting Module Name
    int flag = 0;

    //set bt name
    BT.print(F("AT+NAMEZombieCopter"));
    BT.flush();
    delay(1500);

    flag = print_bt_response();
    if(flag == 1)
    {
        Serial.print(F("Module took new name ... Success\n\n"));
    }
    else if(flag == 0)
    {
        Serial.print(F("Module Name Change Failed\n\n"));
    }
}
```

Summary

In this chapter I covered the using Bluetooth on the Android and the Arduino. I started with an overview of developing programs on the Android and then briefly covered the Java language that is used to create programs on Android devices. I then discussed how I implemented Bluetooth communication on Android with my custom classes. Next, I discussed the Bluetooth adapter that I used to provide the Arduino with Bluetooth communication ability. Finally, I covered the Arduino code that was needed to initialize the Bluetooth adapter.

Chapter 4

Simple Wireless Intruder Alarm System with Motion Detector

This chapter covers a simple wireless intruder alarm system using the Android, Arduino, and a motion detector. The motion detector is covered including the Arduino code needed to get it to work. Next, the Sound class for Android is discussed which creates and manages the sound effects for the Android controller program. Then, I cover the Arduino C code needed for the alarm to function. Finally, I discuss the Android Java code needed for the simple alarm system.

Human Sensor Overview

In this section I cover the device that will detect the motion of an intruder. I first cover the hardware and describe the pin connections needed to get it working. I then discuss the Arduino code that will be needed to get this device to work with the Arduino.

Physical Description

The infrared motion detector used with the projects in this book was purchased from Amazon.com. The exact listing was:

2013newestseller New 5 X HC-SR501 Adjust Ir Pyroelectric Infrared PIR Motion Sensor Detector Modules

and it was bought for $8.49 which was the price for 5 of these motion detectors. This is very inexpensive especially when compared to the price charged by commercial home security alarm companies. However, some units may be bad and other customers have reported that only some of the 5 units were actually working correctly. In my case the first 2 units I tried did not work correctly and would give me positive results even when there was no movement. However, the third unit I tried worked fine. The front of the motion sensor is shown in Figure 4-1.

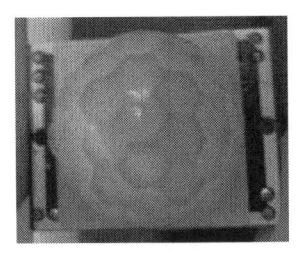

Figure 4-1. HC-SR501 PIR motion detector front with cover

For this particular HC-SR501 sensor the plastic dome shaped cover is removable and reveals the pin labels underneath. See Figure 4-2.

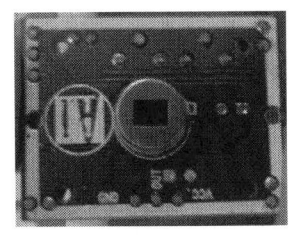

Figure 4-2. HC-SR501 PIR motion detector front without cover

A close up of the pin labels is shown in Figure 4-3.

Figure 4-3. HC-SR501 PIR motion detector front without cover

The pins connect as follows:

- GND - Connect to the GND on the Arduino

- OUT - This pin outputs a 1 if an intruder has been detected and 0 otherwise. It is connected to the pin on the Arduino that is designated as an input pin that will read the voltage level output by this pin.

- VCC – Connect this pin to the 5 volt output pin on the Arduino.

On the underside of the motion sensor you have two controls that allow you to adjust the sensitivity and a time delay adjustment. Try to use the motion detector without touching these adjustments first. I was able to get the motion detector working without adjusting these items. See Figure 4-4.

Figure 4-4. HC-SR501 PIR motion detector back

Arduino Code for the HC-SR501 PIR motion detector

The motion detector is connected to the Arduino and will need to be controlled by code on the Arduino side for our intruder alarm system.

The IRSensorTripped variable holds the status of the motion detector. If the value is true then the motion detector has been tripped and an intruder has been detected. If the value is false then no truder has been detected yet. The variable is declared as a boolean and initialized to false.

boolean IRSensorTripped = false;

The PIN_DETECT variable holds the Arduino pin number that will read in the voltage level from the motion sensor. This variable is declared as an integer and set to pin 8 on the Arduino.

int PIN_DETECT = 8; // digital pin

The SensorValue variable holds the value read in from the motion sensor. It is declared as an integer and initialized to 0 which means no intruder has been detected.

int SensorValue = 0;

The NumberHits variable holds the number of times that an intruder has been detected by the motion sensor. It is declared as an integer and is initialized to 0.

int NumberHits = 0;

The SetupIRProximitySensor() function initializes the motion sensor by setting the PIN_DETECT pin on the Arduino to be an INPUT pin. It does this by calling the pinMode(PIN_DETECT, INPUT) function. See Listing 4-1.

Listing 4-1. The SetupIRProximitySensor() function

```
void SetupIRProximitySensor()
{
```

```
Serial.println(F("Initializing IR Proximity Sensor ...."));

pinMode(PIN_DETECT, INPUT);

}
```

The TestIRProximitySensor() function reads in the value from the motion sensor and returns true if the sensor has detected motion and false if the sensor detects no motion.

Specifically, the function does the following:

1. Reads in the sensor value by calling the digitalRead(PIN_DETECT) function to read the value from pin PIN_DETECT.

2. If the sensor value read in at Step 1 is greater than 0 then increase the total number of intruders detected, print out a message with the total number of intruders, and set the return result to true.

3. Suspend program execution for 25 milliseconds by calling the delay(25) function.

4. Return the result variable result.

See Listing 4-2.

Listing 4-2. The TestIRProximitySensor() function

```
boolean TestIRProximitySensor()
{
    boolean result = false;

    SensorValue = digitalRead(PIN_DETECT); //getting values from IR sensor

    if (SensorValue > 0)
    {
        NumberHits++;

        Serial.print("IR SENSOR(Value,NumberHits)= ");

        Serial.print(SensorValue);//output value to serial display

        Serial.print(" , ");

        Serial.println(NumberHits);

        result = true;
    }
    delay(25);
```

return result;

}

Android Sound Effects

This section discusses the sound effects in the Android program. I first discuss the class that is used for the sound effects which is the Sound class. I then show how this class is used in the MainActivity class to initialize the sounds for the Android program.

The Custom Sound Class

The Sound class is a custom class that is responsible for playing back digital sounds.

The m_SoundPool variable holds a SoundPool class object that was created in the MainActivity class. The SoundPool object is the main interface to loading in, playing back, and managing our sounds.

private SoundPool m_SoundPool;

The m_SoundIndex variable holds the identification number of the sound. The number is the ID of the sound within m_SoundPool.

private int m_SoundIndex = -1;

The m_LeftVolume variable holds a value from 0 no volume to 1 full volume of the left channel of a stereo sound.

float m_LeftVolume = 1;

The m_RightVolume variable holds a value from 0 no volume to 1 full volume of the right channel of a stereo sound.

float m_RightVolume = 1;

The m_Priority variable holds the streaming priority of the sound with 0 being the lowest priority.

int m_Priority = 1;

If the m_Loop variable is set to −1 then the sound is set to loop continuously. If it is 0 then the sound is set to play one time.

int m_Loop = 0;

The m_Rate variable is the speed that the sound is played back. A 1 setting is normal playback speed. The range is from 0.5 to 2.0. The variable is initialized to 1.

float m_Rate = 1;

The m_StreamID variable holds the identification number of the sound stream that is currently playing the actual sound.

int m_StreamID = 0;

The Sound class constructor creates a Sound class object by loading in and initializing the sound designated by the ResourceID input parameter and putting it inside the existing SoundPool class object designated by the Pool input parameter. The Activity class object also must be input as the iContext parameter.

More specifically:

1. The global class variable m_SoundPool is set to the Pool input parameter.

2. The sound is loaded into the m_SoundPool class object by calling the m_SoundPool.load(iContext, ResourceID, 1) function. This function also returns the index number for the newly created sound in the SoundPool object. This index is then assigned to m_SoundIndex.

See Listing 4-3.

Listing 4-3. The Sound constructor

```
Sound(Context iContext, SoundPool Pool, int ResourceID)

{

        m_SoundPool = Pool;

        m_SoundIndex = m_SoundPool.load(iContext, ResourceID, 1);

}
```

The PlaySound() function plays the sound through the Android device's speakers. PlaySound(true) plays the sound in a continuous infinite loop. A PlaySound(false) plays the sound one time only.

More specifically the function does the following:

1. If the m_Loop variable is equal to −1 then this sound is set to continuously play so there is no need to play it again so return without doing anything further.

2. If the input parameter Loop is set to true then set m_Loop to −1 to indicate this sound needs to be looped continuously. Otherwise, set m_Loop to 0.

3. Start playing the sound effect by calling the m_SoundPool.play(m_SoundIndex, m_LeftVolume, m_RightVolume, m_Priority, m_Loop, m_Rate) function. The m_SoundIndex parameter indicates the sound within the sound pool that should be played back. The m_LeftVolume and m_RightVolume indicate the volume levels of the sound's left and right channels. The next parameters indicate the priority, whether to loop the sound or not, and the speed at which to play the sound back. The id number of the stream that is playing back the sound is returned by the function and stored in the m_StreamID class variable.

See Listing 4-4.

Listing 4-4. The PlaySound() function

```
void PlaySound(boolean Loop)
{
    //public final int play (int soundID, float leftVolume, float rightVolume, int priority, int loop, float rate)
    // Sound already playing
    if (m_Loop == -1)
    {
        return;
    }
    if (Loop)
    {
        m_Loop = -1;
    }
    else
    {
        m_Loop = 0;
    }
    m_StreamID = m_SoundPool.play(m_SoundIndex, m_LeftVolume, m_RightVolume, m_Priority, m_Loop, m_Rate);
}
```

The StopSound() function stops the sound playing by:

1. Calling the m_SoundPool.stop(m_StreamID) function with the id number of the stream associated with this playing sound.

2. Setting the m_Loop variable to 0 to indicate that this sound is not to be looped.

See Listing 4-5.

Listing 4-5. The StopSound() function

```
void StopSound()
{
```

```
                m_SoundPool.stop(m_StreamID);

                m_Loop = 0;

}
```

The GetSoundPool() function returns m_SoundPool which is a reference to the sound pool that is being used by this Sound class object. See Listing 4-6.

Listing 4-6. The GetSoundPool() function

```
SoundPool GetSoundPool()

{

                return m_SoundPool;

}
```

Initializing and Using the Sound Class

The Sound class is used in the MainActivity class for such things like playing sounds for indicating an intruder has been detected.

The SoundPool class needs to be imported using the import keyword

```
import android.media.SoundPool;
```

The m_SoundPool variable is declared as a SoundPool class object and is initialized to null.

```
private SoundPool  m_SoundPool = null;
```

Two variables that are references to the Sound class are declared and initialized to null. These variables are m_Alert1SFX and m_Alert2SFX.

```
private Sound  m_Alert1SFX = null;

private Sound m_Alert2SFX = null;
```

The CreateSound() function creates new Sound class objects for use with our intruder alarm system.

The function does the following:

1. Creates a new Sound object using the .wav sound file located in the res\raw folder of the main workspace directory for this project. This is done by calling the Sound constructor with the resource id for this sound file which is R.raw.playershoot2. This sound effect is played when the "Take Photo" button is pressed for the home security system that includes an ArduCAM Mini camera module.

2. Creates a new Sound object using the .wav sound file located in the res\raw folder of the main workspace directory for this project. This is done by calling the Sound constructor with

the resource id for this sound file which is R.raw.explosion1. This sound effect is played when an intruder has been detected or a photo has finished loading in for the home security system that includes an ArduCAM Mini camera module.

> Note: The R class is a globally available built in and automatically generated class that contains resource ids that link to the actual sound files which are wave files located in the res\raw directory of the Android project. You can record your own wave files, copy the files into this directory and change these lines of code to change the sounds being played for functions.

See Listing 4-7.

Listing 4-7. The CreateSound() function

```
void CreateSound(Context iContext)
{
        m_Alert1SFX = new Sound(iContext, m_SoundPool, R.raw.playershoot2);
        m_Alert2SFX = new Sound(iContext, m_SoundPool, R.raw.explosion1);
}
```

The StopSounds() function stops all the sound effects from playing by calling the StopSound() function on each of the Sound class objects. See Listing 4-8.

Listing 4-8. The StopSounds() function

```
void StopSounds()
{
        m_Alert1SFX.StopSound();
        m_Alert2SFX.StopSound();
}
```

The CreateSoundPool() function creates a new SoundPool object by:

1. Calling the SoundPool constructor SoundPool(maxStreams, streamType, srcQuality) with the following input parameters:

    1. maxStreams - The maximum number of simultaneous streams for this SoundPool object. A value of 10 is assigned

    2. streamType - The audio stream type as described in AudioManager. This is set to a value of STREAM_MUSIC.

3. srcQuality - the sample-rate converter quality. Currently has no effect. Set to 0 for the default setting

2. The newly created SoundPool class object is then assigned to m_SoundPool.

See Listing 4-9.

Listing 4-9. The CreateSoundPool() function

```
void CreateSoundPool()
{
    int maxStreams = 10;
    int streamType = AudioManager.STREAM_MUSIC;
    int srcQuality = 0;

    m_SoundPool = new SoundPool(maxStreams, streamType, srcQuality);
    if (m_SoundPool == null)
    {
        Log.e("Main Activity ", "m_SoundPool creation failure!!!!!!!!!!!!!!!!!!!!!!!!!!!!!!!!!!!!!!!!!!!!!!!!!");
    }
}
```

The onCreate() function is a custom function derived from the built in standard Android onCreate() function. The @Override keyword indicates that this function overrides the function defined in the parent class.

The onCreate() function does the following:

1. Creates a new SoundPool object that will hold, maintain, and manipulate the sound effects for this Android program by calling the CreateSoundPool() function.

2. Creates the individual sounds that will be used in this program by calling the CreateSound(this) function.

See Listing 4-10.

Listing 4-10. The onCreate() function

```
@Override
```

```
protected void onCreate(Bundle savedInstanceState)

{

    // Init Sounds

    CreateSoundPool();

    CreateSound(this);

}
```

Arduino Code Overview

This section covers the Arduino code that must be installed in order to get the simple burglar/intruder alarm system working.

The PhoneNumber variable holds the phone number that the Android cell phone will try and call out to when the motion detector has been tripped by an intruder. You will need to change this phone number to the phone number that you want the alarm system to contact when an intruder has been detected.

String PhoneNumber = "9876543210";

The RawCommandLine variable holds the complete command line sent from the Android including any parameters.

String RawCommandLine = "";

The Command variable holds the command sent by the Android without any parameters.

String Command = "";

The SystemActivated variable is set by default to false which means the alarm system is off and will not respond to human movement. A true value means that the alarm system is on and the motion sensor will respond to human movement and notify the Android device that an intruder has been detected.

boolean SystemActivated = false;

The setup() function is the first function called when the Arduino is first turned on or is reset. The setup() function initializes the program by:

1. Initializing the debug Serial Monitor connection by calling the Serial.begin(9600) function with 9600 to set the connection speed to 9600 baud which is the default speed of the Serial Monitor.

2. Prints some key information to the Serial Monitor such as the title of the program running, and the pins on the Arduino serving as the Rx and Tx data receive and transmit pins for Bluetooth.

3. Initializes the Bluetooth. For more detailed information on this please see Chapter 3 in the section discussing Bluetooth on the Arduino.

4. Initializes the motion detector.

5. Displays an initial message with background information to this program.

See Listing 4-11.

Listing 4-11. The setup() function

```
void setup()
{
    // Initialize Serial
    Serial.begin(9600);
    Serial.println(F("Android/Arduino Remote Wireless Intruder Alarm System Version 1.0 for UNO Initializing ... "));
    Serial.println();
    Serial.println();
    Serial.print(F("Software Serial RX, TX Pins are: "));
    Serial.print(RxD);
    Serial.print(F(","));
    Serial.println(TxD);
    Serial.println();
    Serial.println();

    // Initialize Bluetooth
    BT.begin(9600);
    delay(1000);
    Serial.println(F("Initializing Bluetooth ..."));
    initialize_bluetooth();
    delay(1000);

    // Set up the IR Sensor
    SetupIRProximitySensor();

    // Print initial message
```

```
    InitialMessage();

}
```

The ExecuteCommand() function actually executes the command sent by the Android to the Arduino. The command is the input parameter Command and is a String class object.

The ExecuteCommand() function does the following:

1. If the Command is "SystemStart" then turn on the intruder alarm system. Also, send a message back to the Android device confirming that the alarm system has been activated by calling the BT.print(F("SystemStartOK\n")) function which sends the string "SystemStartOK\n" to the Android.

2. If the Command is "SystemStop" then turn off the intruder alarm system. Reset the motion detector by setting the tripped status IRSensorTripped to false. Respond back to the Android device by calling the BT.print(F("SystemStopOK\n")) function which sends the string "SystemStopOK\n" to the Android.

3. If the Command is "GetSystemStatus" then if the intruder alarm system is turned on then send the string "SystemActivated\n" back to Android by calling the BT.print(F("SystemActivated\n")) function. If the intruder alarm system is off then send the string "SystemDeactivated\n" back to the Android device by calling the BT.print(F("SystemDeactivated\n")) function.

4. If the Command is "GetIntruderStatus" then the Arduino will return the status of the motion detector. If the motion detector has been tripped then the BT.print(F("IntruderDetected\n")) function is called in order to send the text string "IntruderDetected\n" as a response. If the motion detector has not been tripped then the BT.print(F("NOIntruderDetected\n")) function is called to send the Android the string "NOIntruderDetected\n".

5. If the Command is "GetPhoneNumber" then the Arduino sends the emergency call out phone number to the Android. The emergency phone number is sent to the Android by calling the BT.print(PhoneNumber) function with the phone number followed by the BT.print(F("\n")) function which sends an end of text data character '\n'.

6. If the Command is not recognized then print out an error message to the Serial Monitor.

See Listing 4-12.

Listing 4-12. The ExecuteCommand() function

```
void ExecuteCommand(String Command)

{

  if (Command == "SystemStart")

  {

    SystemActivated = true;

    BT.print(F("SystemStartOK\n"));
```

```
      Serial.println(F("System is Activated ...."));
    }
    else
    if (Command == "SystemStop")
    {
      SystemActivated = false;
      IRSensorTripped = false;
      BT.print(F("SystemStopOK\n"));
      Serial.println(F("System is DeActivated and IR Sensor is Reset ...."));
    }
    else
    if (Command == "GetSystemStatus")
    {
     if (SystemActivated)
      {
       BT.print(F("SystemActivated\n"));
      }
      else
      {
       BT.print(F("SystemDeactivated\n"));
      }
     }
     else
     if (Command == "GetIntruderStatus")
     {
       if (IRSensorTripped)
        {
          BT.print(F("IntruderDetected\n"));
        }
```

```
    else
    {
      BT.print(F("NOIntruderDetected\n"));
    }
  }
  else
  if (Command == "GetPhoneNumber")
  {
    BT.print(PhoneNumber);
    BT.print(F("\n"));
  }
  else
  {
    Serial.print(F("*************************** The Command: "));
    Serial.print(Command);
    Serial.println(F(" is not recognized **********************************"));
  }
}
```

The ParseCommand() function splits an array of characters based on a user input character and stores the result in an array of strings.

The ParseCommand() function does the following:

1. For each character in the commandline compare that character to the splitcharacter if they are equal then this marks the end of a user parameter so save the temp string in the array of strings Result. If the character is not the splitcharacter then add this character to the temp string variable.

2. Add the temp string variable to the array of strings Result. This adds in the last user parameter that does not have a splitcharacter after it to the array of strings.

3. Returns the command and any user parameters in the Result array.

See Listing 4-13

Listing 4-13. The ParseCommand() function

```
void ParseCommand(const char* commandline, char splitcharacter, String* Result)
{
    int ResultIndex = 0;
    int length = strlen(commandline);
    String temp = "";

    for (int i = 0; i < length ; i++)
    {
        char tempchar = commandline[i];
        if (tempchar == splitcharacter)
        {
            Result[ResultIndex] += temp;
            ResultIndex++;
            temp = "";
        }
        else
        {
            temp += tempchar;
        }
    }
    // Put in end part of string
    Result[ResultIndex] = temp;
}
```

The ParseRawCommand() function separates out commands from parameters. For this example there are no command parameters but we leave this here as a good example of how you can expand and customize this program to suit your needs whatever they may be.

The function does the following:

1. Calls the ParseCommand(RawCommandLine.c_str(), ' ', Entries) function that breaks up the user's input into an array of strings.

2. The first value in the Entries string array is the command and is assigned to Command.

3. If the command matches one of the valid commands then a notice is printed to the Serial Monitor to acknowledge the command.

See Listing 4-14.

Listing 4-14. The ParseRawCommand() function

```
void ParseRawCommand(String RawCommandLine)
{
  String Entries[10];

  // Parse into command and parameters
  ParseCommand(RawCommandLine.c_str(), ' ', Entries);

  Command = Entries[0];

  if (Command == "SystemStart")
  {
    Serial.println(F("Command is SystemStart ...."));
  }
  else
  if (Command == "SystemStop")
  {
    Serial.println(F("Command is SystemStop ...."));
  }
  else
  if (Command == "GetSystemStatus")
  {
    Serial.println(F("Command is GetSystemStatus ...."));
  }
```

```
    else
    if (Command == "GetIntruderStatus")
    {
      Serial.println(F("Command is GetIntruderStatus ..."));
    }
    else
    if (Command == "GetPhoneNumber")
    {
      Serial.println(F("Command is GetPhoneNumber ..."));
    }
}
```

The DisplayCommandParameters() function for this example is empty but serves as a placeholder where parameters can be displayed to the Serial Monitor for debugging purposes. See Listing 4-15.

Listing 4-15. The DisplayCommandParameters() function

```
void DisplayCommandParameters()
{
}
```

The DisplayEndCommandMessage() function prints out the Command that has been executed to the Serial Monitor. See Listing 4-16.

Listing 4-16. The DisplayEndCommandMessage() function

```
void DisplayEndCommandMessage()
{
   Serial.print(F("Command "));
   Serial.print(Command);
   Serial.println(F(" has been executed ..."));


   Serial.println();
```

```
    Serial.println();

    Serial.println();
}
```

The GetCommand() function reads in incoming data from the Bluetooth adapter until there is no more data and returns this data as a String class object.

The function does the following:

1. While there is more data to read in from the Bluetooth adapter

    1. Read in the data one character at a time and add it to the string variable that is to be returned.

2. Return the string read in from Step 1.

See Listing 4-17.

Listing 4-17. The GetCommand() function

```
String GetCommand()
{
    String command = "";

    char out;

    while(BT.available() > 0)
    {
        out = (char)BT.read();

        command += out;
    }
    return command;
}
```

The loop() function is called continuously by the Arduino after the setup() function is called until the Arduino is reset or is shut off and is the main entry point for user created code.

The loop() function does the following:

1. Waits until:

    a. Data become available on the Bluetooth connection, reads in the data, and then assigns it to the RawCommandLine variable which represents the command with any parameters that were sent to the Arduino from the Android device. A break is executed which exits out of the infinite while() loop and continues execution after the while statement.

   b. Or, if the alarm system is turned on and if the motion sensor has not been tripped by an intruder, test the motion sensor. If the motion sensor detects motion then set IRSensorTripped to true and send a motion alert to the Android device by calling the BT.print(F("IR_MOTION_ALERT\n")) function with a "IR_MOTION_ALERT\n" text string.

2. Prints out the raw command line text data obtained in Step 1.

3. Calls the ParseRawCommand(RawCommandLine) function to separate out the command from the command's parameters.

4. Prints out the command and the command's parameters to the Serial Monitor.

5. Executes the command by calling ExecuteCommand(Command) with the Command string obtained from Step 3.

6. Prints out a message to the Serial Monitor that the command has been executed by calling the DisplayEndCommandMessage() function.

See Listing 4-18.

Listing 4-18. The loop() function

```
void loop()
{
    // Android phone as master and Arudino with bluetooth as slave
    // Wait for Command from Android
    while(1)
    {
        // Get Command from Andriod if available
        if(BT.available() > 0)
        {
            // If command is coming in then read it
            RawCommandLine = GetCommand();
            break;
        }
        else
        if (SystemActivated)
```

```
        {
            // Test for Intruder if IR Sensor not already tripped
            if (!IRSensorTripped)
            {
                if (TestIRProximitySensor() == true)
                {
                    IRSensorTripped = true;

                    // Send IR Tripped Message To Android
                    BT.print(F("IR_MOTION_ALERT\n"));
                }
            }
        }
    }


    // Print out the command from Android
    Serial.print(F("Raw Command from Android: "));
    Serial.println(RawCommandLine);


    // Parse Raw Command into Command and Parameters
    ParseRawCommand(RawCommandLine);


    // Print out Command and Parameters
    Serial.println(F("----------------------------------------------------"));
    Serial.print(F("Command: "));
    Serial.println(Command);


    DisplayCommandParameters();
```

```
        Serial.println(F("-----------------------------------------------------"));


        // Execute the command

        ExecuteCommand(Command);


        // Display End of command message

        DisplayEndCommandMessage();

    }
```

Android Code Overview

This section describes the Android program code that will be needed for our simple home intruder alarm to work on the Android side of the alarm system.

MainActivity Class Member Variables

This section describes the class member variables for the MainActivity class.

The m_CallOutActive variable is true if the alarm system is set to call out to the emergency phone number when an intruder is detected. If it is false then the an intruder alert message just appears locally on the debug window of the Android application and a sound is played continuously and a text speech voice is played to announce that an intruder has been detected by the motion detector.

boolean m_CallOutActive = true;

The m_EmergencyPhoneNumber variable holds the phone number for the Android to call if there is an intruder detected by the alarm system. This number is set to null or "" on the Android side and the number itself is read in from the Arduino on application start up.

String   m_EmergencyPhoneNumber = "";

The m_EmergencyMessageRepeatTimes variable holds the number of times to repeat the vocal alarm generated by Android's text to speech function. For example, the default number of times to repeat the phrase "Intruder Alert Intruder has been detected by the motion sensor" is set to 5 times.

int   m_EmergencyMessageRepeatTimes = 5;

The m_SendGetPhoneNumberCommand is true if a command to get the emergency call out phone number from Arduino is to be sent. It is initialized to false.

boolean   m_SendGetPhoneNumberCommand = false;

The REQUEST_ENABLE_BT variable is a user defined value that must be used as a parameter to the startActivityForResult(enableBtIntent, REQUEST_ENABLE_BT) function that is called if the Bluetooth on the Android device is available and turned off. This is the request to display the user prompt that allows the user to select the option of turning on the Bluetooth from within the application instead of going into the Android system settings menu. The default value of this variable is 1.

int   REQUEST_ENABLE_BT = 1;

The m_BlueToothActive variable is true if the Android's Bluetooth is connected to the Arduino and false otherwise.

boolean m_BlueToothActive = false;

The m_BluetoothAdapter represents the Android device's Bluetooth adapter.

BluetoothAdapter m_BluetoothAdapter = null;

The m_UUID is the id is used to connect the Android to the Arduino's Serial Bluetooth adapter. See the Bluetooth section in Chapter 3 for more details.

UUID m_UUID = UUID.fromString("00001101-0000-1000-8000-00805F9B34FB");

The m_Device holds a BluetoothDevice class object.

BluetoothDevice m_Device = null;

The m_DeviceFound variable is true if the Bluetooth adapter on the Arduino is found and false otherwise.

boolean m_DeviceFound = false;

m_Receiver holds the BroadcastReceiver class object that is used when a Bluetooth device is discovered. See the Bluetooth section in Chapter 3 for more information.

BroadcastReceiver m_Receiver;

The m_OutputMsgView represents the Output Window on the Android app that the Bluetooth device name and device address are displayed in.

EditText m_OutputMsgView;

The m_DebugMsgView represents the Debug Window on the Android app that displays the debug information for the application as well as intruder alerts.

EditText m_DebugMsgView;

The m_OutputMsg string holds the text that will be displayed in the Output Window

String m_OutputMsg;

The m_DebugMsg string holds the text that will be displayed in the Debug Message Window.

String m_DebugMsg;

The m_ClientConnectThread is the class object that will be used to establish a Bluetooth connection with the Arduino as well as send commands to the Arduino.

BlueToothClientConnectThread m_ClientConnectThread   = null;

The m_BluetoothMessageHandler variable holds the class object that processes the incoming data from the Arduino that is received using Bluetooth. For example, for Android commands to the Arduino that require a response, this class object will handle it.

BluetoothMessageHandler m_BluetoothMessageHandler = null;

The m_TTS variable is used for text to speech conversion so that the Android device can vocalize such things as intruder alerts and important system messages.

TextToSpeech m_TTS = null;

The m_IR_MOTION_ALERT variable is true if the Arduino's motion detector has been tripped and false otherwise.

boolean m_IR_MOTION_ALERT = false;

The m_StartSystemCommandFinished variable is true if the "StartSystem" Android command has just received a response from the Arduino that needs to be processed.

boolean m_StartSystemCommandFinished = false;

The m_StopSystemCommandFinished variable is true if the "StopSystem" Android command has just received a response from the Arduino that needs to be processed.

boolean m_StopSystemCommandFinished = false;

The m_CameraAvailable variable is false if there is not a camera available which is the case for this example which only uses a motion detector and does not use a camera. We will cover the use of the ArduCAM Mini digital camera in the next chapter.

boolean m_CameraAvailable = false;

The m_ArduinoAlarmSystemON is true if the alarm system is on and false otherwise.

boolean m_ArduinoAlarmSystemON = false;

The m_GetSystemStatusCommandFinished variable is true if the Android's "GetSystemStatus" command has received a response from the Arduino that requires processing.

boolean m_GetSystemStatusCommandFinished = false;

The m_GetIntruderStatusCommandFinished variable is true if the Android "GetIntruderStatus" command has received a response from the Arduino that requires processing.

boolean m_GetIntruderStatusCommandFinished = false;

The m_GetPhoneNumberCommandFinished variable is true if the Android command "GetPhoneNumber" has received a response from the Arduino that needs to be processed.

boolean m_GetPhoneNumberCommandFinished = false;

The MainActivity onCreate() Function

The onCreate() function is called upon the Android application's creation and this function performs the initialization of the simple home intruder alarm program.

The onCreate() function does the following:

1. Executes the code in the parent function located in the parent class by calling super.onCreate(savedInstanceState);

2. Initializes the graphical user interface by calling the setContentView(R.layout.activity_main) function which using the xml code in the res/layout directory to create the Android application's user interface.

3. The orientation of the Android's screen is then set permanently to portrait mode by calling the this.setRequestedOrientation(ActivityInfo.SCREEN_ORIENTATION_PORTRAIT) function with the portrait orientation value. This means that the orientation of the screen will not change when the Android phone is moved from vertical orientation to horizontal orientation.

4. Initializes text to speech by calling the TextToSpeechInit() function.

5. The sound effects are initialized by calling the CreateSoundPool() and CreateSound(this) functions. Refer to the section on Sound in this chapter for more information.

6. A test image is loaded into the window that is reserved for displaying a picture by calling the LoadYUVToImageViewDisplay(R.drawable.tulips_yuyv422_prog_packed_qcif) function. However, since in this example there is no camera attached to the Arduino this just serves as a placeholder for when a camera will be added in the future.

7. A reference to the Output Window is found by calling the findViewById(R.id.outputmsg) function with the resource id that refers to the Output Window EditText.

8. The string "BlueTooth Devices: \n" is added to the m_OutputMsg string which contains the text that will be displayed in the Output Window.

9. The Output Window is then actually updated with this new text by calling the m_OutputMsgView.setText(m_OutputMsg.toCharArray(), 0, m_OutputMsg.length()) function with the character array and the length of the character array as input parameters.

10. Turns off the ability for the user to edit the contents of the Output Window by calling the m_OutputMsgView.setFocusable(false) function which disables the ability of the user to select the window and to change focus to the window.

11. Stops the sound effects for indicating an intruder has been detected by creating a new View.OnClickListener() object, defining the onClick() function in this object to stop the playing of the sound associated with the detection of an intruder, and then setting this object for the Output Window by calling the m_OutputMsgView.setOnClickListener() function. When the user clicks on the Output Window any intruder alert sounds being played will be stopped.

12. Step 11 is repeated for the m_DebugMsgView.

13. The Bluetooth Message Handler is initialized by calling the InitializeMessageHandling() function.

14. Bluetooth is initialized. Refer to the Bluetooth section in Chapter 3 for more detailed information.

See Listing 4-19.

Listing 4-19. The onCreate() function

```java
@Override
protected void onCreate(Bundle savedInstanceState) {
    super.onCreate(savedInstanceState);
    setContentView(R.layout.activity_main);

    this.setRequestedOrientation(ActivityInfo.SCREEN_ORIENTATION_PORTRAIT);

    // Initialize Text to Speech engine
    TextToSpeechInit();

    // Init Sounds
    CreateSoundPool();
    CreateSound(this);

    // YUV Resource File to Bitmap Test
    LoadYUVToImageViewDisplay(R.drawable.tulips_yuyv422_prog_packed_qcif); // Ok

    // Initialize the Output Message Window
    m_OutputMsgView = (EditText) findViewById(R.id.outputmsg);
    m_OutputMsg = "BlueTooth Devices: \n";
    m_OutputMsgView.setText(m_OutputMsg.toCharArray(), 0, m_OutputMsg.length());
    m_OutputMsgView.setFocusable(false);
    m_OutputMsgView.setOnClickListener(new View.OnClickListener()
    {
        public void onClick(View v) {
            m_Alert2SFX.StopSound();
        }
```

```java
        });


// Initialize the Debug Message Window
m_DebugMsgView = (EditText) findViewById(R.id.debugmsg);
m_DebugMsg = "Debug Output: \n";
m_DebugMsgView.setText(m_DebugMsg.toCharArray(), 0, m_DebugMsg.length());
m_DebugMsgView.setFocusable(false);
m_DebugMsgView.setOnClickListener(new View.OnClickListener()
        {
                public void onClick(View v) {
                        m_Alert2SFX.StopSound();
                }


        });


// Initializes the Android to Arduino Message Handling
InitializeMessageHandling();


// Test for Bluetooth Adapter
// If Bluetooth is supported and it is not enabled then ask the user
// if he wants to turn it on.
m_BluetoothAdapter = BluetoothAdapter.getDefaultAdapter();
if (m_BluetoothAdapter == null) {
        // Device does not support Bluetooth
        Log.e("BlueToothTest", "Device does not support Bluetooth!");
        m_DebugMsg += "Device does not support Bluetooth! \n";
        m_DebugMsgView.setText(m_DebugMsg.toCharArray(), 0, m_DebugMsg.length());
}
else
```

```
                {
                        Log.d("BlueToothTest","Device DOES Support Bluetooth!!!!");

                        m_DebugMsg += "Device DOES Support Bluetooth!!!! \n";

                        m_DebugMsgView.setText(m_DebugMsg.toCharArray(), 0, m_DebugMsg.length());


                        // Test if Bluetooth is enabled
                        if (!m_BluetoothAdapter.isEnabled())
                        {
                            Log.d("BlueToothTest","Blue Tooth is NOT Enabled. Request to activate BlueTooth!!!!!!!");

                            m_DebugMsg += "Blue Tooth is NOT Enabled. Request to activate BlueTooth !!!!!!! \n";

                            m_DebugMsgView.setText(m_DebugMsg.toCharArray(), 0, m_DebugMsg.length());


                            Intent enableBtIntent = new Intent(BluetoothAdapter.ACTION_REQUEST_ENABLE);

                            startActivityForResult(enableBtIntent, REQUEST_ENABLE_BT);
                        }
                        else
                        {
                            // Blue Tooth is Enabled so initialize it.
                            InitializeBlueTooth();
                        }
                }
        }
```

The MainActivity TextToSpeechInit() Function

The TextToSpeechInitI() function initializes a new text to speech object variable m_TTS by setting the language to be spoken to English. The function does this by calling the m_TTS.setLanguage(Locale.US) function with the Locale.US value as an input parameter. See Listing 4-20.

Listing 4-20. The TextToSpeechInit() function

```
void TextToSpeechInit()
{
    m_TTS = new TextToSpeech(MainActivity.this, new TextToSpeech.OnInitListener() {
        @Override
        public void onInit(int status)
        {
            if(status == TextToSpeech.SUCCESS)
            {
                int result=m_TTS.setLanguage(Locale.US);
                if (result==TextToSpeech.LANG_MISSING_DATA ||
                    result==TextToSpeech.LANG_NOT_SUPPORTED)
                {
                    Log.e("error", "This Language is not supported");
                }
            }
            else
                Log.e("error", "Initilization Failed!");
        }
    });
}
```

The MainActivity InitializeMessageHandling() Function

The InitializeMessageHandling() function creates a new BluetoothMessageHandler class object and assigns it to m_BluetoothMessageHandler which is a global class member variable. See Listing 4-21.

Listing 4-21. The InitializeMessageHandling() function

```
void InitializeMessageHandling()
{
```

```
                    // Create Test Message Handler

                    m_BluetoothMessageHandler = new BluetoothMessageHandler(this);

}
```

Menu Code

This section discusses the code for generating the Android application's menu which is located in the res/menu directory.

The menu consists of selections for "System Arming" and "Emergency Call Out".

The "System Arming" menu consists of:

- Arm System – This option turns on the intruder alarm system on the Arduino side.

- Disarm System – This option turns off the intruder alarm system on the Arduino side.

- Get Armed Status – This option returns the armed status of the intruder alarm status to the Android.

- Get Motion Detection Status – This option returns the status of the motion detector.

The "Emergency Call Out" menu consists of:

- Enable Emergency Call Out – This option enables the emergency call out to the emergency phone number read in from the Arduino.

- Disable Emergency Call Out – This option disables the emergency call out function. Intruder alerts will generate a sound alert and a voice alert by the Android's text to speech capability.

See Listing 4-22.

Listing 4-22. Android menu XML code

```xml
<menu xmlns:android="http://schemas.android.com/apk/res/android" >


    <item android:id="@+id/SystemArming" android:title="System Arming">

      <menu>

        <item android:id="@+id/arm" android:title="Arm System"/>

        <item android:id="@+id/disarm" android:title="Disarm System"/>
```

```xml
            <item android:id="@+id/getarmedstatus" android:title="Get Armed Status"/>

            <item android:id="@+id/getmotiondetectionstatus" android:title="Get Motion Detection Status"/>

        </menu>

    </item>


    <item android:id="@+id/callout" android:title="Emergency Call Out">

        <menu>

            <item android:id="@+id/calloutenabled" android:title="Enable Emergency Call Out"/>

            <item android:id="@+id/calloutdisabled" android:title="Disable Emergency Call Out"/>

        </menu>

    </item>

</menu>
```

The MainActivity onOptionsItemSelected() Function

The onOptionsItemSelected() function is called whenever a menu item is selected by the user.

The onOptionsItemSelected() function does the following:

1. If the user selects the "Arm System" menu selection then the SendSystemArmCommmandToArduino(true) function is called with the value true to send a command to the Arduino to turn on the alarm system.

2. If the user selects the "Disarm System" menu selection then the SendSystemArmCommmandToArduino(false) function is called with the value of false to send a command to the Arduino to turn off the alarm system.

3. If the user selects the "Get Armed Status" menu selection then the SendGetStatusCommandToArduino() function is called to send a command to the Arduino to get the activation status of the alarm and a notification of whether the alarm system is turned on or off is returned.

4. If the user selects the "Get Motion Detection Status" menu selection then the SendGetIntruderStatusCommandToArduino() function is called to send a command to the Arduino to get the tripped status of the motion detector and the status of the motion detector is returned.

5. If the user selects the "Enable Emergency Call Out" menu item then the SetCallOutStatus(true) function with the true parameter is called to enable the Android's emergency call out function.

6. If the user selects the "Disable Emergency Call Out" menu item then the SetCallOutStatus(false) function is called to disable the Android's emergency call out function.

See Listing 4-23.

Listing 4-23. The onOptionsItemSelected() function

```java
@Override
public boolean onOptionsItemSelected(MenuItem item) {
    // Handle item selection
    switch (item.getItemId()) {
        case R.id.arm:
            SendSystemArmCommmandToArduino(true);
            return true;

        case R.id.disarm:
            SendSystemArmCommmandToArduino(false);
            return true;

        case R.id.getarmedstatus:
            SendGetStatusCommandToArduino();
            return true;

        case R.id.getmotiondetectionstatus:
            SendGetIntruderStatusCommandToArduino();
            return true;

        case R.id.calloutenabled:
            SetCallOutStatus(true);
            return true;

        case R.id.calloutdisabled:
            SetCallOutStatus(false);
```

```
            return true;


      default:

          return super.onOptionsItemSelected(item);

   }
}
```

The MainActivity SendSystemArmCommmandToArduino() Function

The SendSystemArmCommmandToArduino() turns the alarm system on or off by sending an Arm/Disarm command to the Arduino depending on the value of the input parameter.

The SendSystemArmCommmandToArduino() function does the following:

1. If the Active input parameter is true then turn on the alarm system. The text to speech m_TTS vocalizes the words "Arming System" by calling the m_TTS.speak() function. The command to send to the Arduino is set to "SystemStart".

2. If the Active input parameter is false then turn off the alarm system. The text to speech m_TTS vocalizes the words "Disarming System" by calling the m_TTS.speak() function. The command to send to the Arduino is set to "SystemStop".

3. The command determined in Steps 1 and 2 is set in the Bluetooth message handler by calling the m_BluetoothMessageHandler.SetCommand(Command) function with the command value.

4. Next, if there is a good Bluetooth connection between the Android and the Arduino then write out the command to the Arduino by calling the m_ClientConnectThread.GetConnectedThread().write(Command.getBytes()) function with the command string from Steps 1 and 2 converted into an array of bytes as the input parameter.

5. Also, the m_DebugMsg string variable is updated with debug information to be displayed in the Debug Window. The m_DebugMsgView.setText() function actually updates the Debug Window's text to reflect that latest changes in the m_DebugMsg string.

See Listing 4-24.

Listing 4-24. The SendSystemArmCommmandToArduino() function

```
void SendSystemArmCommmandToArduino(boolean Active)

{

            String Command = "";
```

```java
// Send Command To Arduino to Arm/Disarm system
if (Active)
{
        m_TTS.speak("Arming System", TextToSpeech.QUEUE_ADD, null);
        Command = "SystemStart";
}
else
{
        m_TTS.speak("Disarming System", TextToSpeech.QUEUE_ADD, null);
        Command = "SystemStop";
}
m_DebugMsg += "Sending System command to Arduino!! \n";


// Set Up Data Handler
m_BluetoothMessageHandler.SetCommand(Command);
if (m_ClientConnectThread != null)
{
    if (m_ClientConnectThread.GetConnectedThread() != null)
    {
            m_ClientConnectThread.GetConnectedThread().write(Command.getBytes());
            m_DebugMsg += "Writing System Start/Stop command to Arduino!! \n";
    }
    else
    {
            m_DebugMsg += "ConnectedThread is Null!! \n";
    }
}
else
```

```
            {
                    m_DebugMsg += "ClientConnectThread is Null!! \n";
            }

            m_DebugMsgView.setText(m_DebugMsg.toCharArray(), 0, m_DebugMsg.length());
}
```

The MainActivity SendGetStatusCommandToArduino() Function

The SendGetStatusCommandToArduino() function requests the activation status of the intruder alarm from the Arduino and specifically does the following:

1. Uses Android's text to speech feature to announce that it is getting the status of the alarm system.

2. The command to be sent out to the Arduino is set to "GetSystemStatus".

3. The command is set inside the Bluetooth connection handler by calling the m_BluetoothMessageHandler.SetCommand(Command) function with the command that is being executed and that a return value is expected from the Arduino.

4. If the Bluetooth connection between the Android and Arduino is valid and active then write the command to the Arduino by calling the m_ClientConnectThread.GetConnectedThread().write(Command.getBytes()) function with the command string from Step 2 converted to a byte array.

5. Debug messages are also printed out to the Debug Window on the Android user interface.

See Listing 4-25.

Listing 4-25. The SendGetStatusCommandToArduino() function

```
void SendGetStatusCommandToArduino()
{
            // Send Command To Arduino to Get director of Arudino's SD Card

            m_TTS.speak("Getting Status of Alarm System from Arduino", TextToSpeech.QUEUE_ADD, null);


            m_DebugMsg += "Sending GetSystemStatus command to Arduino!! \n";

            String Command = "GetSystemStatus";
```

```
                m_BluetoothMessageHandler.SetCommand(Command);

            if (m_ClientConnectThread != null)
            {
                if (m_ClientConnectThread.GetConnectedThread() != null)
                {
                    m_ClientConnectThread.GetConnectedThread().write(Command.getBytes());
                    m_DebugMsg += "Writing GetStatus command to Arduino!! \n";
                }
                else
                {
                    m_DebugMsg += "ConnectedThread is Null!! \n";
                }
            }
            else
            {
                m_DebugMsg += "ClientConnectThread is Null!! \n";
            }
            m_DebugMsgView.setText(m_DebugMsg.toCharArray(), 0, m_DebugMsg.length());
}
```

The MainActivity SendGetIntruderStatusCommandToArduino() function

The SendGetIntruderStatusCommandToArduino() function sends the command to the Arduino to retrieve the tripped status of the motion detector.

The function does the following:

1. If the alarm system is not on then notify the user that the intruder status cannot be retrieved since the alarm system is not active then exit the function.

2. If the alarm system is on then the command is set to "GetIntruderStatus".

3. Next, the command is set in the Bluetooth message handler by calling the m_BluetoothMessageHandler.SetCommand(Command) function with the command in Step 2 as the input parameter.

4. The command is actually sent using Bluetooth by calling the m_ClientConnectThread.GetConnectedThread().write(Command.getBytes()) function with the command converted to an array of bytes.

5. Debug messages and error notifications are also displayed in the Debug Window if needed.

See Listing 4-26.

Listing 4-26. The SendGetIntruderStatusCommandToArduino() function

```
void SendGetIntruderStatusCommandToArduino()
{
    if (m_ArduinoAlarmSystemON == false)
    {
        m_TTS.speak("Can not get intruder status because Alarm System is not Active", TextToSpeech.QUEUE_ADD, null);
        return;
    }

    // Send Command To Arduino to Get director of Arudino's SD Card
    m_TTS.speak("Getting Intruder Detection Status from Arduino", TextToSpeech.QUEUE_ADD, null);

    m_DebugMsg += "Sending Intruder Detection command to Arduino!! \n";
    String Command = "GetIntruderStatus";

    m_BluetoothMessageHandler.SetCommand(Command);
    if (m_ClientConnectThread != null)
    {
        if (m_ClientConnectThread.GetConnectedThread() != null)
        {
            m_ClientConnectThread.GetConnectedThread().write(Command.getBytes());
            m_DebugMsg += "Writing Get Intruder Status command to Arduino!! \n";
        }
```

```
                else
                {
                        m_DebugMsg += "ConnectedThread is Null!! \n";
                }
        }
        else
        {
                m_DebugMsg += "ClientConnectThread is Null!! \n";
        }


        m_DebugMsgView.setText(m_DebugMsg.toCharArray(), 0, m_DebugMsg.length());
}
```

The MainActivity SetCallOutStatus() Function

The SetCallOutStatus() function does the following:

1. Activates the emergency cell phone call out if the input parameter status is true. Also gives a vocalized notification that the call out has been activated.

2. Deactivates the emergency cell phone call out if the input parameter status is false. Also gives a vocalized notification that the call out has been turned off.

See Listing 4-27.

Listing 4-27. The SetCallOutStatus() function

```
void SetCallOutStatus(boolean status)
{
        if (status == true)
        {
                m_CallOutActive = true;
                m_TTS.speak("Activating Emergency Call Out", TextToSpeech.QUEUE_ADD, null);
        }
        else
```

```
            {

                    m_CallOutActive = false;

                    m_TTS.speak("DeActivating Emergency Call Out", TextToSpeech.QUEUE_ADD, null);

            }

}
```

The BluetoothMessageHandler Class

The BluetoothMessageHandler class reads the data from the Arduino sent using Bluetooth in response to commands sent by the Android device. This class also sets up the MainActivity class thread to process the data that is received.

The ReceiveMessage() function is called whenever there is incoming data from the Arduino that is sent by Bluetooth. The data may be part of the response or the entire response and this class will handle either case and make sure that all the data is read in.

The ReceiveMessage() function does the following:

1. If the Android command waiting for a response from the Arduino is "SystemStart" then the ProcessSystemStartCommand() function is called to process this incoming data.

2. If the Android command waiting for a response from the Arduino is "SystemStop" then the ProcessSystemStopCommand() function is called to process this incoming data.

3. If the Android command waiting for a response from the Arduino is "GetSystemStatus" then the ProcessGetStatusCommand() function is called to process this incoming data.

4. If the Android command waiting for a response from the Arduino is "GetIntruderStatus" then the ProcessGetIntruderStatusCommand() function is called to process this incoming data.

5. If the Android command waiting for a response from the Arduino is "GetPhoneNumber" then the ProcessGetPhoneNumberCommand() function is called to process this incoming data.

6. If the command set in the m_Command string does not match any of the above commands then issue an error notice by calling the Log.e() function.

See Listing 4-28.

Listing 4-28. The ReceiveMessage() function

```
void ReceiveMessage(int NumberBytes, byte[] Message)

{

            // Process Incoming Data

            // Assume incoming data is associated with the current m_Command variable
```

```
            // and process the Message accordingly.

            if (m_Command == "SystemStart")

            {

                    // System is monitoring for Intruder Alerts.

                    ProcessSystemStartCommand(NumberBytes,Message);

            }

            else

            if (m_Command == "SystemStop")

            {

                    ProcessSystemStopCommand(NumberBytes,Message);

            }

            else

            if (m_Command == "GetSystemStatus")

            {

                    ProcessGetStatusCommand(NumberBytes,Message);

            }

            else

            if (m_Command == "GetIntruderStatus")

            {

                    ProcessGetIntruderStatusCommand(NumberBytes, Message);

            }

            else

            if (m_Command == "GetPhoneNumber")

            {

                    ProcessGetPhoneNumberCommand(NumberBytes, Message);

            }

            else

            {

                    Log.e("BlueToothTest" , "Error - Command for Data Receive Not Found!!!!!!");
```

            }
}

The ProcessSystemStartCommand() function processes data from the Arduino that responds to the activation of the intruder alarm by:

1. Calling the ReceiveTextData() function to process the incoming text data.

2. If the data has finished loading in then:

    1. Notify the MainActivity class that there is a response from Arduino to process by calling the m_MainActivity.SetStartSystemCommandFinished() function.

    2. Update the MainActivity class thread to process the newly read in data by calling the m_MainActivity.runOnUiThread(m_MainActivity) function which will execute the run() function in the MainActivity class.

See Listing 4-29.

Listing 4-29. The ProcessSystemStartCommand() function

```
void ProcessSystemStartCommand(int NumberBytes, byte[] Message)

{

                boolean FinishedReceivingText = ReceiveTextData(NumberBytes, Message);

                if (FinishedReceivingText)

                {

                        // Process SystemStart Command

                        m_MainActivity.SetStartSystemCommandFinished();

                        Log.d("BlueTooth Test","Start System Command has Finished ... ");


                        // Update the User Interface

                        m_MainActivity.runOnUiThread(m_MainActivity);

                }

}
```

The ProcessSystemStopCommand() function processes Bluetooth data from the system stop command by:

1. Receiving the text data response from Arduino by calling the ReceiveTextData() function.

2. After receiving all of the data call the m_MainActivity.SetStopSystemCommandFinished() function to notify the MainActivity class that this data needs to be processed.

3. Calling the m_MainActivity.runOnUiThread(m_MainActivity) to actually process this new incoming data from the Arduino.

See Listing 4-30.

Listing 4-30. The ProcessSystemStopCommand() function

```
void ProcessSystemStopCommand(int NumberBytes, byte[] Message)
{
    boolean FinishedReceivingText = ReceiveTextData(NumberBytes, Message);
    if (FinishedReceivingText)
    {
        // Process SystemStart Command
        m_MainActivity.SetStopSystemCommandFinished();
        Log.d("BlueTooth Test","Stop System Command has Finished ... ");

        // Update the User Interface
        m_MainActivity.runOnUiThread(m_MainActivity);
    }
}
```

The ProcessGetStatusCommand() function gets the activation status of the alarm system by doing the following:

1. Receives the text data response from the Arduino by calling the ReceiveTextData() function.

2. After the data has finished being transmitted the m_MainActivity.SetGetSystemStatusCommandFinished() function is called to indicate that this data now needs to be processed.

3. The text data is then actually processed by calling the m_MainActivity.runOnUiThread(m_MainActivity) function which executes the run() function in the MainActivity class.

See Listing 4-31.

Listing 4-31. The ProcessGetStatusCommand() function

```
void ProcessGetStatusCommand(int NumberBytes, byte[] Message)
{
        boolean FinishedReceivingText = ReceiveTextData(NumberBytes, Message);
        if (FinishedReceivingText)
        {
                // Process Remote Directory Command
                m_MainActivity.SetGetSystemStatusCommandFinished();
                Log.d("BlueTooth Test","Get Alarm System Status has Returned ... ");


                // Update the User Interface
                m_MainActivity.runOnUiThread(m_MainActivity);
        }
}
```

The ProcessGetIntruderStatusCommand() function gets the motion detection status of the alarm by:

1. Reading in the text data from the Arduino by calling the ReceiveTextData() function.

2. If the text data has finished being received then call the m_MainActivity.SetGetIntruderStatusCommandFinished() function to tell the MainActivity class that the data is ready to be processed.

3. The data then is actually processed by calling the m_MainActivity.runOnUiThread(m_MainActivity) function which executes the run() function in the MainActivity class.

See Listing 4-32.

Listing 4-32. The ProcessGetIntruderStatusCommand() function

```
void ProcessGetIntruderStatusCommand(int NumberBytes, byte[] Message)
{
        boolean FinishedReceivingText = ReceiveTextData(NumberBytes, Message);


        if (FinishedReceivingText)
```

```
        {
                // Process Remote Directory Command

                m_MainActivity.SetGetIntruderStatusCommandFinished();

                Log.d("BlueTooth Test","Get Intruder Status has Returned ... ");


                // Update the User Interface

                m_MainActivity.runOnUiThread(m_MainActivity);

        }
}
```

The ProcessGetPhoneNumberCommand() function gets the emergency phone number that is being sent from the Arduino by:

1. Receiving text data by calling the ReceiveTextData() function.

2. When the data has finished being read, notifying the MainActivity class that the data is ready to be processed by calling the m_MainActivity.SetGetPhoneNumberCommandFinished() function.

3. Processing the actual text data by calling the m_MainActivity.runOnUiThread(m_MainActivity) function which executes the run() function in the MainActivity class.

See Listing 4-33.

Listing 4-33. The ProcessGetPhoneNumberCommand() function

```
void ProcessGetPhoneNumberCommand(int NumberBytes, byte[] Message)
{
        boolean FinishedReceivingText = ReceiveTextData(NumberBytes, Message);

        if (FinishedReceivingText)
        {
                // Process Remote Directory Command

                m_MainActivity.SetGetPhoneNumberCommandFinished();

                Log.d("BlueTooth Test","Get Phone Number Command has Returned ... ");


                // Update the User Interface

                m_MainActivity.runOnUiThread(m_MainActivity);
```

                }
        }

The MainActivity Run() Function

The run() function in the MainActivity class is called from the BluetoothMessageHandler class to process the incoming data from Arduino and to update the Android's user interface.

The run() function does the following:

1. If a response to the Start System command has been received then call the ProcessStartSystemCommandResult() function.

2. If a response to the Stop System command has been received then call the ProcessStopSystemCommandResult() function.

3. If a command to retrieve the emergency phone number needs to be sent to Arduino then call the SendGetPhoneNumberMessage() function.

4. If a response to the Get Emergency Phone number command has been received then call the ProcessGetPhoneNumberCommand() function.

5. If a response to the Get System Status command has been received then call the ProcessGetSystemStatusValue() function.

6. If a response to the Get Intruder Status command has been received then call the ProcessGetIntruderStatusValue() function.

See Listing 4-34.

Listing 4-34. The Run() function

```
// Updates data in User Interface

public void run()

{
                // SystemStart Command Callback

                if (m_StartSystemCommandFinished)

                {
                        ProcessStartSystemCommandResult();

                        m_StartSystemCommandFinished = false;

                }
```

```
// SystemStop Command Callback.
if (m_StopSystemCommandFinished)
{
        ProcessStopSystemCommandResult();
        m_StopSystemCommandFinished = false;
}


// Get Phone Number From Arduino
if (m_SendGetPhoneNumberCommand)
{
        SendGetPhoneNumberMessage();
        m_SendGetPhoneNumberCommand = false;
}


// GetPhoneNumber Command Callback
if (m_GetPhoneNumberCommandFinished)
{
        ProcessGetPhoneNumberCommand();
        m_GetPhoneNumberCommandFinished = false;
}


// GetSystemStatus Command Callback
if (m_GetSystemStatusCommandFinished)
{
        // System has returned the status of the Alarm system on the
        // Arduino side
        ProcessGetSystemStatusValue();
        m_GetSystemStatusCommandFinished = false;
}
```

```
                // Get Intruder Status Command Callback

                if (m_GetIntruderStatusCommandFinished)

                {

                        ProcessGetIntruderStatusValue();

                        m_GetIntruderStatusCommandFinished = false;

                }

}
```

The ProcessStartSystemCommandResult() function processes the Start System response from the Arduino.

The function does the following:

1. Gets the text data string form the Bluetooth message handler object by calling the m_BluetoothMessageHandler.GetStringData() function.

2. Trims the newline '\n' character and other white space characters from the string by calling the trim() function.

3. If the text data is "SystemStartOK" then indicate that the alarm is now active by a voice message using the text to speech function of the Android and a text message to the Debug Window on the Android application. Also the m_ArduinoAlarmSystemON is set to true to indicate on the Android side that the alarm system is now active.

4. If the text data is "IR_MOTION_ALERT" then an intruder has been detected by the motion sensor attached to the Arduino. If the emergency call out feature of the system is activated then the CallEmergencyPhoneNumber() function is called. Otherwise the text to speech engine is used to vocalize an intruder alert, a text message is written to the Debug Window and a sound effect is continuously played until the user clicks on either the Output Window or the Debug Window.

5. If the text data is neither "SystemStartOK" nor "IR_MOTION_ALERT" then there is an error condition. The error is vocalized and printed to the Debug Window.

See Listing 4-35.

Listing 4-35. The ProcessStartSystemCommandResult() function

```
void ProcessStartSystemCommandResult()

{

        String Data = m_BluetoothMessageHandler.GetStringData();

        String DataTrimmed = Data.trim();
```

```java
if (DataTrimmed.equalsIgnoreCase("SystemStartOK"))
{
    m_ArduinoAlarmSystemON = true;
    m_TTS.speak("System Has Started on Arduino Side", TextToSpeech.QUEUE_ADD, null);
    m_DebugMsg = "";
    AddDebugMessage("System Has Started on Arduino Side");
}
else
if (DataTrimmed.equalsIgnoreCase("IR_MOTION_ALERT"))
{
    m_IR_MOTION_ALERT = true;
    m_DebugMsg = "";
    AddDebugMessage("INTRUDER DETECTED!!!!");

    if (m_CallOutActive)
    {
        CallEmergencyPhoneNumber();
    }
    else
    {
        // Repeat message
        for (int i = 0; i < m_EmergencyMessageRepeatTimes ; i++)
        {
            m_TTS.speak("Intruder Alert Intruder has been detected by the motion sensor", TextToSpeech.QUEUE_ADD, null);
        }
        m_Alert2SFX.PlaySound(true);
    }
}
else
```

```
        {
                m_TTS.speak("ERROR in System Start command response from Arduino",
TextToSpeech.QUEUE_ADD, null);

                AddDebugMessage("ERROR in System Start command response from Arduino!!!!");

        }
}
```

The ProcessStopSystemCommandResult() function processes the return result from the Stop System command by:

1. Retrieving text data from the Bluetooth message handler class.

2. The text data is trimmed of extra whitespace characters including the newline character.

3. The text data is compared to "SystemStopOK" if it matches then the alarm system is set to off on the Android side, a vocalized notification is spoken, and a text message is displayed on the Debug Window.

4. If the text data does not match "SystemStopOK" then generate a vocalized error and display an error message in the Debug Window.

See Listing 4-36.

Listing 4-36. The ProcessStopSystemCommandResult() function

```
void ProcessStopSystemCommandResult()
{
        String Data = m_BluetoothMessageHandler.GetStringData();

        String DataTrimmed = Data.trim();

        if (DataTrimmed.equalsIgnoreCase("SystemStopOK"))
        {
                m_ArduinoAlarmSystemON = false;

                m_TTS.speak("System Has Stopped on Arduino Side", TextToSpeech.QUEUE_ADD, null);

                m_DebugMsg = "";

                AddDebugMessage("System Has Stopped on Arduino Side");
        }
        else
        {
```

```
            m_TTS.speak("ERROR in System Stop command response from Arduino",
TextToSpeech.QUEUE_ADD, null);

                    AddDebugMessage("ERROR in System Stop command response from Arduino!!!!");

        }

}
```

The SendGetPhoneNumberMessage() function sends a GetPhoneNumber command to the Arduino using Bluetooth by doing the following:

1. Vocalizing the action by using the Android's text to speech feature.

2. Setting the command inside the Bluetooth message handler class to "GetPhoneNumber".

3. If the Bluetooth connection between the Android device and the Arduino is active then send the command "GetPhoneNumber" to the Arduino using Bluetooth.

See Listing 4-37.

Listing 4-37. The SendGetPhoneNumberMessage() function

```
void SendGetPhoneNumberMessage()

{

        // Send Command To Arduino to Get director of Arudino's SD Card

        m_TTS.speak("Getting Emergency Phone Number from Arduino", TextToSpeech.QUEUE_ADD, null);


        m_DebugMsg += "Sending Get Emergency Phone Number command to Arduino!! \n";

        String Command = "GetPhoneNumber";


        m_BluetoothMessageHandler.SetCommand(Command);

        if (m_ClientConnectThread != null)

        {

                if (m_ClientConnectThread.GetConnectedThread() != null)

                {

                        m_ClientConnectThread.GetConnectedThread().write(Command.getBytes());

                        m_DebugMsg += "Writing GetPhoneNumber command to Arduino!! \n";

                }
```

```
                else
                {
                        m_DebugMsg += "ConnectedThread is Null!! \n";
                }
        }
        else
        {
                m_DebugMsg += "ClientConnectThread is Null!! \n";
        }
        m_DebugMsgView.setText(m_DebugMsg.toCharArray(), 0, m_DebugMsg.length());
}
```

The ProcessGetPhoneNumberCommand() function processes the response to the GetPhoneNumber command by:

1. Retrieving the text data from the message handler class by calling the m_BluetoothMessageHandler.GetStringData() function.

2. Trimming the newline end of data character from the end of the text data and assigning the value to m_EmergencyPhoneNumber.

3. Vocalizing the phone number and printing the number to the Debug Window.

See Listing 4-38.

Listing 4-38. The ProcessGetPhoneNumberCommand() function

```
void ProcessGetPhoneNumberCommand()
{
        String Data = m_BluetoothMessageHandler.GetStringData();

        String DataTrimmed = Data.trim();


        m_EmergencyPhoneNumber = DataTrimmed;

        m_TTS.speak("Emergency call out phone number is " + m_EmergencyPhoneNumber, TextToSpeech.QUEUE_ADD, null);


        m_DebugMsg = "";
```

AddDebugMessage("Emergency call out phone number is " + m_EmergencyPhoneNumber + "\n");

}

The ProcessGetSystemStatusValue() function processes the response to the Get System Status command from the Arduino by:

1. Getting the text data from the Bluetooth message handler class.

2. Trimming the newline character from the end of the text data as well as any other whitespace characters.

3. If the text data is equal to "SystemActivated" then the alarm system is active and running so give a vocalized announcement and then set the command in the bluetooth message handler to "SystemStart" to wait for an intruder alert message since the system is active.

4. If the text data is equal to "SystemDeActivated" then vocalize an announcement.

5. If the text data does not match one of the above then vocalize an error announcement and print out an error to the Android error console by calling the Log.e() function.

See Listing 4-39.

Listing 4-39. The ProcessGetSystemStatusValue() function

```
void ProcessGetSystemStatusValue()
{
    String Data = m_BluetoothMessageHandler.GetStringData();
    String DataTrimmed = Data.trim();

    if (DataTrimmed.equalsIgnoreCase("SystemActivated"))
    {
        m_ArduinoAlarmSystemON = true;
        m_TTS.speak("Alarm System is running", TextToSpeech.QUEUE_ADD, null);

        // Since system has been activated we need to switch back the bluetooth
        // handler command to SystemStart to receive intruder alerts.
        m_BluetoothMessageHandler.SetCommand("SystemStart");
    }
    else
```

```java
            if (DataTrimmed.equalsIgnoreCase("SystemDeActivated"))
            {
                m_ArduinoAlarmSystemON = false;
                m_TTS.speak("Alarm System is not running", TextToSpeech.QUEUE_ADD, null);
            }
            else
            {
                m_TTS.speak("ERROR in result from get alarm system status command", TextToSpeech.QUEUE_ADD, null);
                Log.e("MAINACTVITY","DataTrimmed GetSystemStatus Value = " + "\"" + DataTrimmed + "\"");
            }
}
```

The ProcessGetIntruderStatusValue() function processes the return value from the Get Intruder Status command by:

1. Retrieving the text data from the Bluetooth message handler.

2. Trimming the text data of whitespace characters from the end.

3. If the text data is equal to "IntruderDetected" then a vocalized announcement is spoken and a message is printed to the Debug Output window.

4. If the text data is equal to "NOIntruderDetected" then a vocalized announcement is spoken.

5. If the text data is not one of the above then an error announcement is made and an error is printed out using the Log.e() function.

6. If alarm system has been activated we need to switch back the Bluetooth handler command to "SystemStart" in order to wait for further intruder alerts and notifications

See Listing 4-40.

Listing 4-40. The ProcessGetIntruderStatusValue() function

```java
void ProcessGetIntruderStatusValue()
{
    String Data = m_BluetoothMessageHandler.GetStringData();
    String DataTrimmed = Data.trim();
```

```
            if (DataTrimmed.equalsIgnoreCase("IntruderDetected"))

            {

                    m_IR_MOTION_ALERT = true;

                    m_DebugMsg = "";

                    AddDebugMessage("INTRUDER DETECTED!!!!");

                    m_TTS.speak("Intruder Detected", TextToSpeech.QUEUE_ADD, null);

            }

            else

            if (DataTrimmed.equalsIgnoreCase("NOIntruderDetected"))

            {

                    m_IR_MOTION_ALERT = false;

                    m_TTS.speak("No intruder detected", TextToSpeech.QUEUE_ADD, null);

            }

            else

            {

                    m_TTS.speak("ERROR ", TextToSpeech.QUEUE_ADD, null);

                    Log.e("MAINACTIVITY", "ERROR, GetIntruderStatus Value is not known = " +
DataTrimmed);

            }


            // If alarm system has been activated we need to switch back the bluetooth

            // handler command to SystemStart in order to wait for further alerts

            if (m_ArduinoAlarmSystemON)

            {

                    m_BluetoothMessageHandler.SetCommand("SystemStart");

            }

}
```

The CallEmergencyPhoneNumber() function actually makes the call by the Android cell phone to the emergency phone number by:

1. Creating a string consisting of "tel:" and the emergency phone number without any trailing newline or other white space characters such as tab, space, etc.

2. Creating a new Intent class object which is set to call out to a phone number.

3. Setting the phone number in the intent object to the emergency phone number from Step 1 by calling the setData() function.

4. Dialing this phone number by calling the startActivity(intent) function with the intent from Step 2 as an input parameter.

See Listing 4-41.

Listing 4-41. The CallEmergencyPhoneNumber() function

```
void CallEmergencyPhoneNumber()
{
    // Call emergency number so that home owner can listen in to what tripped the
    // intruder alarm
    String uri = "tel:" + m_EmergencyPhoneNumber.trim();
    Intent intent = new Intent(Intent.ACTION_CALL);
    intent.setData(Uri.parse(uri));
    startActivity(intent);
}
```

The GUI XML Code

The Graphical User Interface or GUI for short is created by xml code and is located in the res/layout directory.

For the example in this chapter the important items are:

1. The Output Window which is a EditText object with a resource id of outputmsg.

2. The Debug Window which is a EditText object with a resource id of debugmsg.

See Listing 4-42.

> Note: There are references to other GUI items which are not used here but will be used in the hands on example involving the ArduCAM camera in the next chapter.

Listing 4-42. GUI xml code

```xml
<RelativeLayout xmlns:android="http://schemas.android.com/apk/res/android"
    xmlns:tools="http://schemas.android.com/tools"
    android:layout_width="match_parent"
    android:layout_height="match_parent"
    tools:context=".MainActivity" >

    <EditText
        android:id="@+id/outputmsg"
        android:layout_width="wrap_content"
        android:layout_height="wrap_content"
        android:layout_alignParentLeft="true"
        android:layout_below="@+id/imageView1"
        android:layout_toLeftOf="@+id/TestMessageButton"
        android:ems="10"
        android:lines="3"
        android:maxLines="3"
        android:text="@string/hello_world" />

    <EditText
        android:id="@+id/debugmsg"
        android:layout_width="wrap_content"
        android:layout_height="wrap_content"
        android:layout_alignParentLeft="true"
        android:layout_alignRight="@+id/outputmsg"
        android:layout_below="@+id/outputmsg"
        android:ems="10"
        android:lines="4"
        android:maxLines="4"
```

android:text="@string/Debug_Message" />

</RelativeLayout>

The AndroidManifest.xml File

The AndroidManifest.xml file will need to be changed to add permissions for certain functions.

In order to allow a program to call out to a phone number by itself you need to add the CALL_PHONE permission.

<uses-permission android:name="android.permission.CALL_PHONE" />

In order to allow an Android application to use Bluetooth you need to add the following permissions.

<uses-permission android:name="android.permission.BLUETOOTH" />

<uses-permission android:name="android.permission.BLUETOOTH_ADMIN" />

See Listing 4-43.

Listing 4-43. The AndroidManifest.xml file

<?xml version="1.0" encoding="utf-8"?>

<manifest xmlns:android="http://schemas.android.com/apk/res/android"

  package="com.example.bluetoothtest"

  android:versionCode="1"

  android:versionName="1.0" >

  <uses-sdk

    android:minSdkVersion="8"

    android:targetSdkVersion="16" />

  <application

    android:allowBackup="true"

    android:icon="@drawable/ic_launcher"

    android:label="@string/app_name"

```xml
            android:theme="@style/AppTheme" >
        <activity
            android:name="com.example.bluetoothtest.MainActivity"
            android:label="@string/app_name" >
            <intent-filter>
                <action android:name="android.intent.action.MAIN" />

                <category android:name="android.intent.category.LAUNCHER" />
            </intent-filter>
        </activity>
    </application>


    <uses-permission android:name="android.permission.CALL_PHONE" />
    <uses-permission android:name="android.permission.BLUETOOTH" />
    <uses-permission android:name="android.permission.BLUETOOTH_ADMIN" />


</manifest>
```

Summary

In this chapter I discussed a simple intruder alarm system that uses an Android, Arduino, and an infrared motion sensor to detect intruders. I started off by covering the motion detector and the Arduino code needed to get it to operate. Next, I covered the custom made class for Android called the Sound class that is used to manage all the sound effects in the Android application. Then, I discussed the Arduino code needed for the simple intruder alarm system. Finally, I covered the Android code required for the alarm system.

Chapter 5

Hands on Example: Creating a Simple Intruder Alarm System

In this chapter I give a quick start guide to setting up the simple intruder alarm system I discussed in detail in Chapter 4. This is the chapter you want to read for quickly setting up the simple wireless intruder alarm system on the Arduino side and correctly using the software on the Android side to control the alarm system. Understanding this chapter requires no technical expertise and is designed for everyone.

Intruder Alarm System Overview

The general set up of our simple intruder alarm system consists of an Android device communicating with an Arduino equipped with a Bluetooth Adapter and an infrared motion detector. The Android sends commands to the Arduino and the Arduino responds to these commands with text data or binary data. For the system in this chapter the Arduino will only respond with text data. When an intruder is detected a text message will be sent to the Android device. The Android device will then either call out to an emergency phone number or issue an audible alert so that the home owner can be notified that the alarm system has detected an intruder. See Figure 5-1.

Master/Client   Slave/Server

BlueTooth Connection — BlueTooth Adapter

Android Sends Command →

Arduino Responds to Command ←

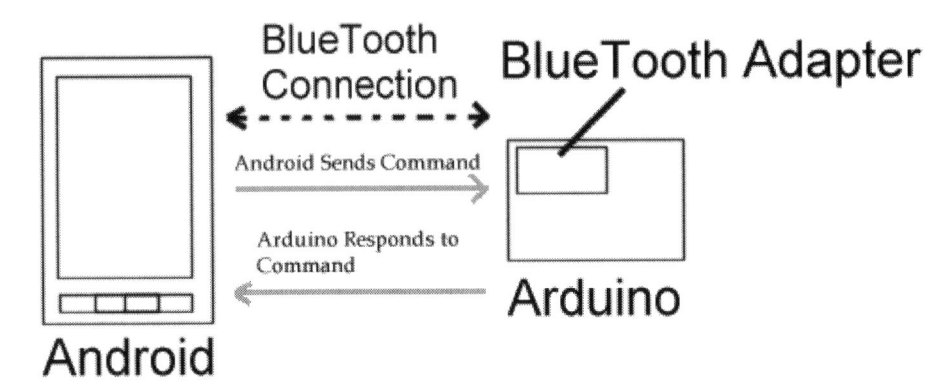

Android   Arduino

Figure 5-1. System overview

Wiring Diagram

In order to create the alarm system in this chapter you will need to connect the components of the system which are:

- An HC-SR501 Pyroelectric Infrared (PIR) Motion Sensor Detector for Arduino
- An HC-06 4 pin Serial Bluetooth Adapter Module for Arduino
- An Arduino UNO
- Jumper wires

See Figure 5-2.

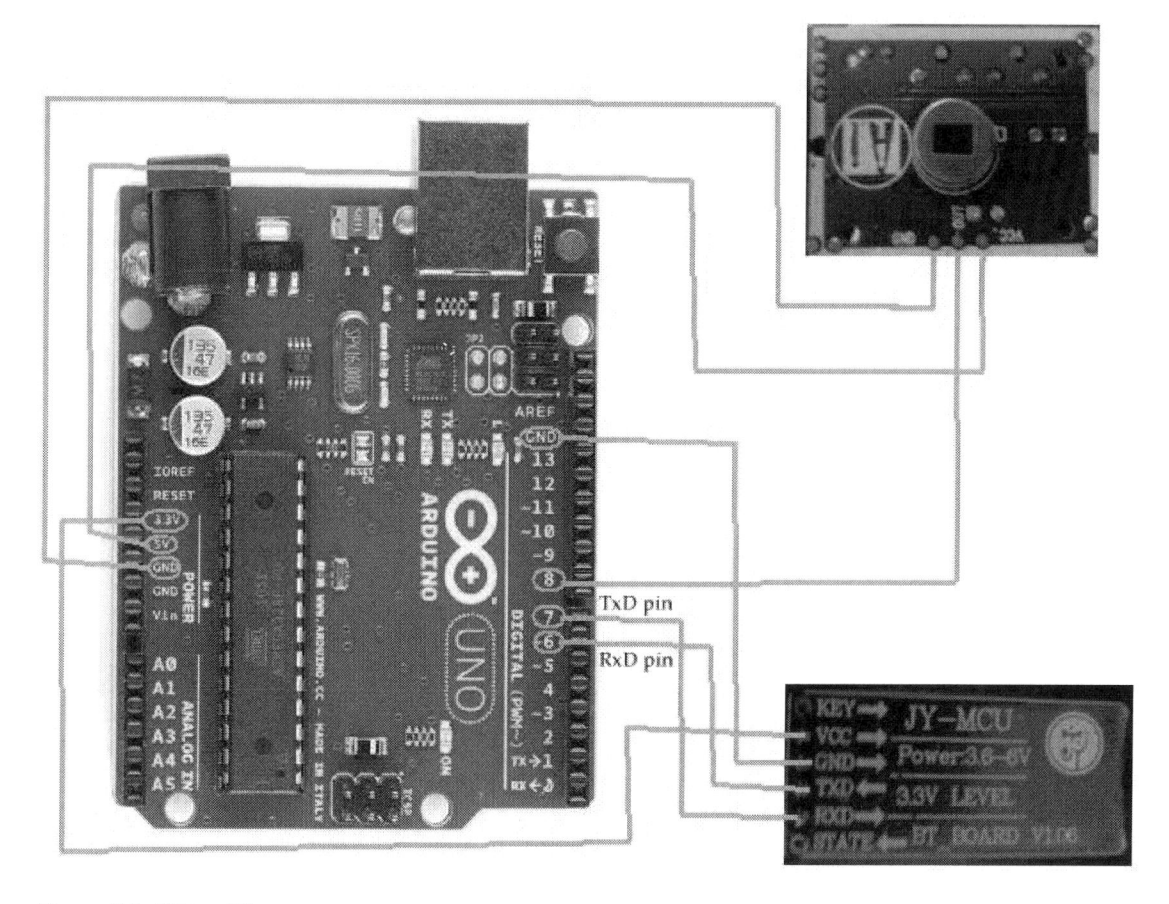

Figure 5-2. Wiring Diagram

Detailed Connection Instructions

To connect the motion sensor to the Arduino UNO you need to:

- Connect the GND pin on the sensor to a GND pin on the Arduino.
- Connect the OUT pin on the sensor to pin 8 on the Arduino.
- Connect the VCC pin on the sensor to the 5 volt pin on the Arduino

To connect the Bluetooth adapter to the Arduino UNO you need to:

- Connect the VCC pin on the sensor to the 3.3 volt pin on the Arduino.
- Connect the GND pin on the sensor to the GND pin on the Arduino.
- Connect the TXD pin on the sensor to pin 6 on the Arduino.
- Connect the RXD pin on the sensor to pin 7 on the Arduino.

Official Support Web Site

The official support web site for this book can be found at:

http://www.psycho-sphere.com/diy.html

Downloading the Final Android APK installable

The final Android executable which is an .apk file can be downloaded directly at:

http://www.psycho-sphere.com/SimpleUNOBurglarAlarm.apk

Copy this file onto your Android device and run it to begin the installation. Usually your Android device has a file manager where you can begin installing an Android .apk file by clicking on it. Also, you may need to allow installation from unknown sources or non-android market sources to install this program to your device. This option is under the Android's system settings.

After installation you should see a new icon representing this program on your Android device.

Downloading the Code

The Arduino Uno source code for this simple intruder alarm is located at:

http://www.psycho-sphere.com/AndroidArduinoIntruderAlarmSystem_UNO.zip

The Android source code for this simple intruder alarm is located at:

http://www.psycho-sphere.com/WorkSpaceUNOIntruderAlarm.zip

Importing Android Code

If you want to install the program from the source code then you will first need to unzip the file using a program such as 7 zip (which is free) and then import the code.

Make sure your Android device is connected in "Developer" mode using your USB cable to your Android development system.

If you are using the ADT bundle you will need to select File->Import->Android->Existing Android Code Into Workspace to get started importing the code. Follow the directions in the subsequent window dialogs to import the code into your current workspace.

After you import the code you can then select "MainActivity" in the leftmost window pane in the Integrated Development Environment or IDE and select Run->Run from the menu to begin installation of the program on your Android device. In the next window popup select "Android Application" to run the program as an Android application.

If you are using Android Studio you will have to first convert the old Eclipse ADT project code files into the new form. Please refer to the latest information at:

http://developer.android.com/sdk/installing/migrate.html

Quick Start User's Guide

1. Connect the Arduino UNO, motion detector, and Bluetooth adapter as shown in Figure 5-2.

2. Make sure you have the Arduino Integrated Development Environment installed on your computer system.

3. Download and unzip the Arduino source code using a program such as 7-zip (which is free at http://www.7-zip.org).

4. Open the Arduino IDE and load in the source code for the simple intruder alarm system.

5. Change the emergency phone number in the Arduino source code to the number you want your Android cell phone to call out to when the intruder alarm system detects an intruder.

   String PhoneNumber = "9876543210"; ← Change this number to your own emergency number

6. Save your customized program.

7. Compile and upload this program to your Arduino UNO that you have connected to your computer using your USB cable. See Chapter 1 for more information on setting up the Arduino and compiling and uploading Arduino programs.

8. Next, install the Android application for the simple intruder alarm system by either downloading the .apk file and installing it or downloading the source code, importing it into your Android development system and then installing it to your Android device.

9. Start up the Serial Monitor on the Arduino IDE which brings up the Serial Monitor window where debug messages from the Arduino program are printed out.

10. You should see the following be displayed in the Serial Monitor window. Please note that the Bluetooth adapter should return a value of "OK" upon successful initialization.

> Note: The code assumes that the Bluetooth Adapter is set for a speed of 9600 baud. If the speed is different from this value then you will need to change the Arduino code. See Chapter 3 for more information.

Android/Arduino Remote Wireless Intruder Alarm System Version 1.0 for UNO Initializing ...

Software Serial RX, TX Pins are: 6,7

Initializing Bluetooth ...

OK

AT Command Test ... Success

Initializing IR Proximity Sensor ....

==

Android/Arduino Remote Intruder Alarm Wireless System Version 1.0 for UNO

Copyright 2015 by Robert Chin. All Rights Reserved.

This code to accompany the book entitled:

Home Security Systems DIY using Android and Arduino

Please refer to the book for detailed explainations of this code

==

11. Next, start up the Android application. If your Bluetooth is not currently activated then the application will attempt to start it up. You will see a popup window requesting permission to turn on Bluetooth. Click the "Yes" button to start up bluetooth. Also, make sure "Airplane Mode" is off before you try to turn on Bluetooth.

12. After clicking "Yes" you should see a message that the Android device is trying to turn on Bluetooth.

13. Next, the application will start up discovery for new Bluetooth enabled devices. When it finds the Arduino's Bluetooth adapter it will issue a pair request if the devices have not been previously paired.

14. Click the text field and enter the password which is generally "1234" and then click "Done" and then "OK". You may need to click on the "SYM" key first to activate the numerical keys.

15. A message will be displayed saying that the devices are now paired and now a Bluetooth connection is being established between the Android device and the Arduino. After the connection is made you should hear vocal announcements as well as text updates in the top Output Window and the bottom Debug Window.

16. On the Arduino side you should see the debug output showing that the Android is requesting and then receiving the emergency phone number from the Arduino.

Raw Command from Android: GetPhoneNumber

Command is GetPhoneNumber ...

Command: GetPhoneNumber

Command GetPhoneNumber has been executed ...

17. Next, we need to arm the system so that it will begin to detect intruders. To do this we press the menu button on the Android device and select the "System Arming" sub menu from the main menu.

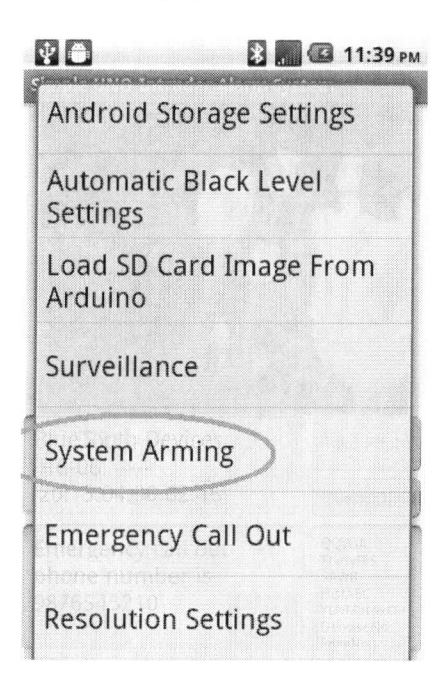

18. Then, from the System Arming menu we need to select "Arm System" to activate the alarm

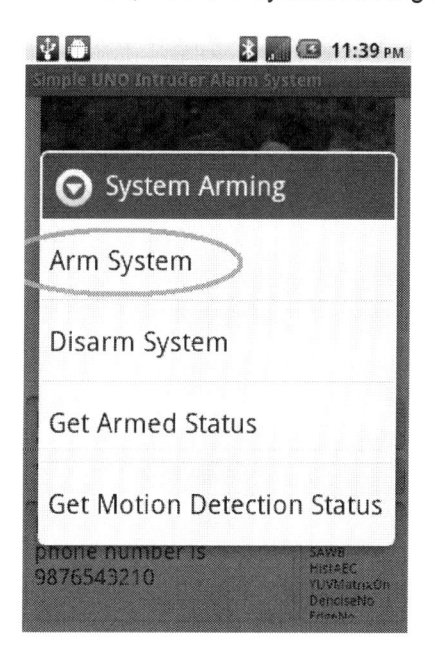

19. On the Arduino side we see that the Android has sent the "SystemStart" command to turn on the intruder alarm system and to activate the motion sensor.

Raw Command from Android: SystemStart

Command is SystemStart ....

Command: SystemStart

System is Activated ....

Command SystemStart has been executed ...

20. Make sure you have the motion sensor set up so that it only detects motion you want it to for example, I used the plastic casing of a box of dental floss that was cut in half to hold the motion sensor.

21. Now, its time to test the intruder alarm. Pass your hand in front of the motion sensor. You should see a message from the Arduino. The following message indicates that the value of the sensor is 1 and the total number of intrusions detected since this Arduino program has started running is 1.

IR SENSOR(Value,NumberHits)= 1 , 1

22. On the Android side by default the call out to the emergency number is triggered by the motion alert message from the Arduino. The Android will attempt to call out to the number specified in the Arduino code as the emergency contact number.

23. Next, click on "End" to end the call.

24. Disarm the system by going back to the "System Arming" menu and selecting the "Disarm System" menu selection. You should also see the Arduino side receiving this "SystemStop" message and disarming the intruder alarm system.

Raw Command from Android: SystemStop

Command is SystemStop ....

Command: SystemStop

System is DeActivated and IR Sensor is Reset ....

Command SystemStop has been executed ...

25. Now, go back to the menu and select the "Emergency Call Out" submenu. Turn off the emergency call out feature by selecting the "Disable Emergency Call Out" option.

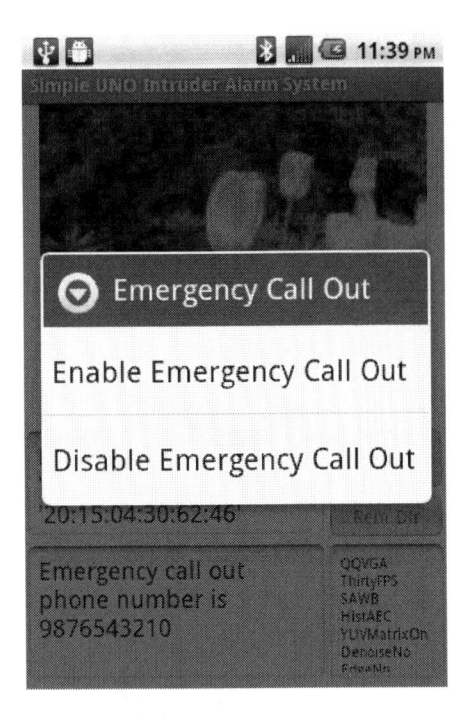

26. Arm the system again and wait for the response from the Arduino acknowledging the command.

27. Trip the motion sensor and this time a vocal announcement and a message in the Debug Window should indicate that an intruder has been detected. Click the Debug Message Window or Output Window to stop the alert sound effect that continuously repeats.

28. The Serial Monitor should also show an updated count of the total number of intruders detected while the Arduino has been powered up or after the last system reset. Number of intruders should be 2.

IR SENSOR(Value,NumberHits)= 1 , 2

29. Finally, disarm the system to reset the motion detector.

30. Finally, arm the system again and your new alarm system is ready to use.

Chapter 6

ArduCAM Mini Wireless Intruder Alarm/Video Surveillance System

This chapter covers the ArduCAM based intruder alarm system. I start by giving an overview of the type of ArduCAM camera that will be used for this project and then how to install the needed software libraries. Next, I cover the additional C/C++ Arduino code needed to support the ArduCAM camera on the Arduino side. Finally, I discuss the additional Java Android code needed to support the ArduCAM camera on the Android side.

ArduCAM Mini Camera Hardware Overview

The ArduCAM Mini ov2640 2 Mega Pixel digital camera that we will use for this project is made by ArduCAM and the official web site is located at:

http://www.arducam.com

The ArduCAM Mini is an easy to use camera that uses the standard SPI and I2C interfaces for the Arduino. The pins on the ArduCAM Mini are as follows:

- CS – "Chip Select" or "Slave Select" which is the pin on each device that the master can use to enable and disable specific devices. When a device's Slave Select pin is low, it communicates with the master. When it's high, it ignores the master. This allows you to have multiple SPI devices sharing the same MISO, MOSI, and CLK lines.

- MOSI – "Master Out Slave In" which is the master line for sending data to the peripherals.

- MISO – "Master In Slave Out" which is the slave line for sending data to the master

- SCK – "Serial Clock" which is the clock pulses which synchronize data transmission generated by the master

- GND – Ground pin.

- VCC – Input Voltage pin that accepts either 3.3 volts or 5.0 volts.
- SDA – 'Data Line" for the I2C interface.
- SCL – "Clock Line" for the I2C interface

See Figure 6-1.

Figure 6-1. ArduCAM Mini ov2640 2 Mega Pixel camera

The ArduCAM Mini ov2640 2 Mega Pixels camera can be purchased in many places including on Amazon.

ArduCAM Mini Camera Library Software Installation

In order to use the ArduCAM Mini Camera with your Arduino you will need to download and install the ArduCAM libraries. You will need to go to the ArduCAM web site listed in the previous section and download the library.

Once you download the zip file you will need to uncompress it using a program like 7-zip and install the two directories "ArduCAM" and "UTFT4ArduCAM_SPI" under the "libraries" directory for Arduino. For example, on my Windows XP system I have installed the ArduCAM libraries in my "Program Files/Arduino/libraries" directory by copying the two directories to this "libraries" directory. After doing this you should be able to compile source code that includes the ArduCAM library.

Arduino Code Overview

This section will cover the Arduino code that specifically relates to using the ArduCAM Mini camera in our intruder alarm system. Some code such as the code that covers the motion detector and Bluetooth communication has been covered in previous chapters so I won't repeat it here.

In order to use the ArduCAM library with Arduino we need to include the two libraries that we added to the Arduino's "libraries" directory in the previous section.

#include <ArduCAM.h>

#include <UTFT_SPI.h>

The libraries needed for the I2C library needs to be included. This is needed for the ArduCAM Mini

#include <Wire.h>

The ArduCAM's SPI interface needs to include the SPI library.

#include <SPI.h>

The "memorysaver.h" include file will save Arduino memory by only including camera register data for the specific cameras defined in the include file.

#include "memorysaver.h"

The CS or chip select pin on the Arduino that controls whether the camera is "live" or active on the SPI bus or not is set to pin 4. This pin is connected to the CS pin on the camera.

// set pin 4 as the slave select for SPI:

const int SPI_CS = 4;

The ArduCAM camera is represented by the myCAM variable which is an ArduCAM class object. The myCAM variable is initialized by a constructor with two input parameters which are the type of camera which is "OV2640" and the chip select pin number on the Arduino.

ArduCAM myCAM(OV2640, SPI_CS);

The camera resolution types are defined as VGA, QVGA, and QQVGA, or None.

enum ResolutionType

{

  VGA,

  QVGA,

  QQVGA,

  None

};

The current resolution of the camera is held in the global variable Resolution and has the default setting of None.

ResolutionType Resolution = None;

The RawCommandLine holds the raw text command string including any parameters sent to the Arduino from the Android device and is initialized to the null string "".

String RawCommandLine = "";

The Command variable holds the text string that represents the command only without any parameters and is initialized to the null string "".

String Command = "";

The PhoneNumber variable holds the text string of the emergency phone number that will be called when the motion detector is tripped.

String PhoneNumber = "9876543210";

The InitializeCamera() function initializes the ArduCAM Mini camera by:

1. Starting up the I2C interface by calling the Wire.begin() or Wire1.begin() functions.

2. Setting the chip select pin number on the Arduino to be an OUTPUT pin by calling the pinMode(SPI_CS, OUTPUT) function with the pin number SPI_CS and the type of pin which is OUTPUT. This outputs a voltage to the CS pin on the camera.

3. Initializes the SPI bus on the Arduino by calling the SPI.begin() function.

4. Test to see if the SPI bus is working by writing the hex value 0x55 to the register ARDUCHIP_TEST1 and then reading the value from the ARDUCHIP_TEST1 register. If the returned value is hex 0x55 then the SPI bus is working correctly. Otherwise, print out an error to the Serial Monitor and go into an infinite loop to halt program execution.

5. Test to see if the ov2640 camera is detected.

6. Set the image format that the ArduCAM Mini camera generates to JPEG by calling the myCAM.set_format(JPEG) function with the input parameter JPEG.

7. Perform further camera initialization by calling the myCAM.InitCAM() function.

See Listing 6-1.

Listing 6-1. The InitializeCamera() function

```
void InitializeCamera()

{
        uint8_t vid, pid;

        uint8_t temp;
```

```
#if defined (__AVR__)
    Wire.begin();
#endif
#if defined(__arm__)
    Wire1.begin();
#endif

Serial.println("ArduCAM Starting ..........");
// set the SPI_CS as an output:
pinMode(SPI_CS, OUTPUT);

// initialize SPI:
SPI.begin();

//Check if the ArduCAM SPI bus is OK
myCAM.write_reg(ARDUCHIP_TEST1, 0x55);
temp = myCAM.read_reg(ARDUCHIP_TEST1);
if (temp != 0x55)
{
    Serial.println("SPI interface Error!");
    while (1);
}

//Check if the camera module type is OV2640
myCAM.rdSensorReg8_8(OV2640_CHIPID_HIGH, &vid);
myCAM.rdSensorReg8_8(OV2640_CHIPID_LOW, &pid);
if ((vid != 0x26) || (pid != 0x42))
    Serial.println("Can't find OV2640 module!");
else
```

Serial.println("OV2640 detected ....");

//Change to BMP capture mode and initialize the OV2640 module

myCAM.set_format(JPEG);

myCAM.InitCAM();

}

The setup() function initializes the intruder alarm system by:

1. Initializing the Serial Monitor and setting the communication rate to 9600 baud by calling the Serial.begin(9600) function.

2. Displaying the title of the program by calling the DisplayTitle() function

3. Initializing the camera by calling the InitializeCamera() function.

4. Initializing the Arduino's Bluetooth adapter by calling the InitializeBlueTooth() function.

5. Initializing the motion detector by calling the SetupIRProximitySensor() function.

6. Printing out an information message by calling the InitialMessage() function.

See Listing 6-2.

Listing 6-2. The setup() function

```
void setup()
{
  Serial.begin(9600);


  DisplayTitle();

  InitializeCamera();

  InitializeBlueTooth();


  // Set up Infrared Motion Sensor

  SetupIRProximitySensor();

  InitialMessage();

}
```

The CaptureImage() function captures a camera image to the camera's on board FIFO memory by doing the following:

1. Resetting the FIFO read pointer to 0 which is the beginning of the image frame by calling the myCAM.flush_fifo() function. This sets the camera to start writing a new image at the beginning of the camera's frame buffer memory or FIFO memory.

2. Resets the FIFO image capture flag by calling the myCAM.clear_fifo_flag() function. This flag is set every time an image has been captured and written to the FIFO buffer. We need to clear this flag before capturing another image.

3. Starts the image capture by calling the myCAM.start_capture() function.

4. Waits until the image capture is done and then continues execution of the program.

See Listing 6-3.

Listing 6-3. The CaptureImage() function

```
void CaptureImage()
{
    // Capture Image from camera to camera's FIFO memory buffer
    Serial.println(F("............. Starting Image Capture ..............."));

    //Flush the FIFO
    myCAM.flush_fifo();

    //Clear the capture done flag
    myCAM.clear_fifo_flag();

    //Start capture
    myCAM.start_capture();
    Serial.println("Start Capture");

    // Wait till capture is finished
    while (!myCAM.get_bit(ARDUCHIP_TRIG , CAP_DONE_MASK));
    Serial.println("Capture Done!");
```

}

The read_fifo_burst() function reads the image data that is stored in the camera's FIFO memory and sends it to the Android using Bluetooth.

The function does this by:

1. Reading in the length of the image stored in the FIFO memory buffer by calling the myCAM.read_fifo_length() function.

2. If the length of the image is greater or equal to 393,216 bytes which is greater than the size of the entire FIFO memory buffer then the end of the image has not been found and an error condition is returned.

3. The camera is selected to be active on the SPI bus by calling the myCAM.CS_LOW() function which sets the CS

4. Setting the camera into FIFO burst mode which speeds up reading from the camera's memory by calling the myCAM.set_fifo_burst() function.

5. Reading in and discarding the first byte of the image which is a dummy byte that does not contain any image information by calling the SPI.transfer(0x00) function.

6. For each byte in the image buffer read the byte by calling the SPI.transfer(0x00) function with the parameter 0x00 and transmit this byte to the Android device using Bluetooth by calling the BT.write(temp) function with temp being the byte to transmit. Delay program execution 10 microseconds between read operations.

7. Disconnect the camera from the SPI bus after the image is read and transmitted to the Android device by calling the myCAM.CS_HIGH() function.

8. Print out the total number of bytes transmitted and the total time it takes for transmission to the Serial Monitor.

See Listing 6-4.

Listing 6-4. The read_fifo_burst() function

```
uint8_t read_fifo_burst()

{

  uint32_t length = 0;

  uint8_t temp,temp_last;


  long bytecount = 0;

  int total_time = 0;
```

```
length = myCAM.read_fifo_length();
if(length >= 393216)  //384 kb
{
  Serial.println("Not found the end.");
   return 0;
}


Serial.print(F("READ_FIFO_LENGTH() = "));
Serial.println(length);


total_time = millis();


myCAM.CS_LOW();
myCAM.set_fifo_burst();
temp = SPI.transfer(0x00); // Added in to original source code, read in dummy byte


length--;
while( length-- )
{
        temp_last = temp;
        temp = SPI.transfer(0x00);


        // Transmit over Bluetooth
        //Serial.write(temp);
        BT.write(temp);
        bytecount++;


        // Commented out because it causes loss of data in transfer
        // so that data does not match amount of data read from read fifo length command
```

```
            //if( (temp == 0xD9) && (temp_last == 0xFF) )

            //       break;


            delayMicroseconds(10);

    }

    myCAM.CS_HIGH();


    Serial.print(F("ByteCount = "));

    Serial.println(bytecount);


    total_time = millis() - total_time;

    Serial.print("Total time used:");

    Serial.print(total_time/1000, DEC);

    Serial.println(" seconds ...");

}
```

The TransmitImageSize() function transmits the size of the image that is captured in the camera's FIFO frame buffer to the Android device using Bluetooth.

The function does this by:

1. Reading the raw length of the image by calling the myCAM.read_fifo_length() function.

2. Calculating the actual size of the image by subtracting 1 from the returned value from step 1.

3. Transmit the size of the captured image to the Android device by calling the BT.print() function with the size of the image and then a text data termination character of '\n'.

See Listing 6-5.

Listing 6-5. The TransmitImageSize() function

```
void TransmitImageSize()

{

        // Transmit Size

        uint32_t FIFOImageLength = myCAM.read_fifo_length();
```

```
        long Size = FIFOImageLength - 1; // Account for 1 dummy byte at beginning of image

        Serial.print(F("FIFO RAW IMAGE SIZE = "));
        Serial.println(FIFOImageLength);

        Serial.print(F("Transmitting Image Length = "));
        Serial.println(Size);

        // Send length of captured image to Android
        BT.print(Size);
        BT.print(F("\n"));
}
```

The SetCameraResolution() function changes the resolution of the camera based on Android commands by doing the following:

1. If the resolution of the image to be captured is QQVGA then set the output JPEG size to 160x120.

2. If the resolution of the image to be captured is QVGA then set the output JPEG size to 320x240.

3. If the resolution of the image to be captured is VGA then set the output JPEG size to 640x480.

4. If the resolution of the image is none of the above then print out an error to the Serial Monitor.

See Listing 6-6.

Listing 6-6. The SetCameraResolution() function

```
void SetCameraResolution()
{
        // Set Screen Size
        if (Resolution == QQVGA)
        {
                myCAM.OV2640_set_JPEG_size(OV2640_160x120);
```

```
            }

            else

            if (Resolution == QVGA)

            {

                    myCAM.OV2640_set_JPEG_size(OV2640_320x240);

            }

            else

            if (Resolution == VGA)

            {

                    myCAM.OV2640_set_JPEG_size(OV2640_640x480);

            }

            else

            {

                    Serial.println(F("ERROR in setting Camera Resolution"));

            }

}
```

The ExecuteCommand() function executes the actual commands coming from the Android device by doing the following:

1. If the command is VGA then set the image capture resolution to VGA. Then set the camera resolution by calling the SetCameraResolution() function. Take a photo using the camera by calling the CaptureImage() function. Finally transmit the image size of the captured photo to the Android device using Bluetooth by calling the TransmitImageSize() function.

2. If the command is QVGA then set the image capture resolution to QVGA. Then set the camera resolution by calling the SetCameraResolution() function. Take a photo using the camera by calling the CaptureImage() function. Finally transmit the image size of the captured photo to the Android device using Bluetooth by calling the TransmitImageSize() function.

3. If the command is QQVGA then set the image capture resolution to QQVGA. Then set the camera resolution by calling the SetCameraResolution() function. Take a photo using the camera by calling the CaptureImage() function. Finally transmit the image size of the captured photo to the Android device using Bluetooth by calling the TransmitImageSize() function.

4. If the command is "GetImageData" then read the image data from the FIFO memory buffer and transmit the data to the Android device using Bluetooth by calling the read_fifo_burst() function.

5. If the Command is "SystemStart" then turn on the intruder alarm system. Also, send a message back to the Android device confirming that the alarm system has been activated by calling the BT.print(F("SystemStartOK\n")) function which sends the string "SystemStartOK\n" to the Android.

6. If the Command is "SystemStop" then turn off the intruder alarm system. Reset the motion detector by setting the tripped status IRSensorTripped to false. Respond back to the Android device by calling the BT.print(F("SystemStopOK\n")) function which sends the string "SystemStopOK\n" to the Android.

7. If the Command is "GetSystemStatus" then if the intruder alarm system is turned on then send the string "SystemActivated\n" back to Android by calling the BT.print(F("SystemActivated\n")) function. If the intruder alarm system is off then send the string "SystemDeactivated\n" back to the Android device by calling the BT.print(F("SystemDeactivated\n")) function.

8. If the Command is "GetIntruderStatus" then the Arduino will return the status of the motion detector. If the motion detector has been tripped then the BT.print(F("IntruderDetected\n")) function is called in order to send the text string "IntruderDetected\n" as a response. If the motion detector has not been tripped then the BT.print(F("NOIntruderDetected\n")) function is called to send the Android the string "NOIntruderDetected\n".

9. If the Command is "GetPhoneNumber" then the Arduino sends the emergency call out phone number to the Android. The emergency phone number is sent to the Android by calling the BT.print(PhoneNumber) function with the phone number followed by the BT.print(F("\n")) function which sends an end of text data character '\n'.

10. If the Command is not recognized then print out an error message to the Serial Monitor.

See Listing 6-7.

Listing 6-7. The ExecuteCommand() function

boolean ExecuteCommand(String Command)

{

    boolean result = true;

    if (Command == "VGA")

    {

        Resolution = VGA;

        SetCameraResolution();

        CaptureImage();

        TransmitImageSize();

    }

```
else
if (Command == "QVGA")
{
        Resolution = QVGA;
        SetCameraResolution();
        CaptureImage();
        TransmitImageSize();
}
else
if (Command == "QQVGA")
{
        Resolution = QQVGA;
        SetCameraResolution();
        CaptureImage();
        TransmitImageSize();
}
else
if (Command == "GetImageData")
{
        read_fifo_burst();
}
else
if (Command == "SystemStart")
{
        SystemActivated = true;
        BT.print(F("SystemStartOK\n"));
        Serial.println(F("System is Activated ...."));
}
else
```

```
if (Command == "SystemStop")

{

        SystemActivated = false;

        IRSensorTripped = false;

        BT.print(F("SystemStopOK\n"));

        Serial.println(F("System is DeActivated and IR Sensor is Reset ...."));

}

else

if (Command == "GetSystemStatus")

{

        if (SystemActivated)

        {

                BT.print(F("SystemActivated\n"));

        }

        else

        {

                BT.print(F("SystemDeactivated\n"));

        }

}

else

if (Command == "GetIntruderStatus")

{

        if (IRSensorTripped)

        {

                BT.print(F("IntruderDetected\n"));

        }

        else

        {

                BT.print(F("NOIntruderDetected\n"));
```

```
                }
        }
        else
        if (Command == "GetPhoneNumber")
        {
                BT.print(PhoneNumber);
                BT.print(F("\n"));
        }
        else
        {
                Serial.print(F("Command '"));
                Serial.print(Command);
                Serial.println(F("' is not recognized ..."));
                result = false;
        }
        return result;
}
```

> Note: The loop() function which is continuously executed by the Arduino is functionally the same as in the loop() function discussed for the simple intruder alarm system. Please refer to the chapter on the simple intruder alarm system for a more detailed discussion of the loop() function.

Android Code Overview

In this section we will cover the Android code that is specific to the ArduCAM version of the simple intruder alarm system described in previous chapters.

The following variables are from the MainActivity class.

The m_InfoTextView TextView class object holds resolution information for the photo that the user wants taken. The choices are QQVGA, QVGA, and VGA.

```
TextView m_InfoTextView;
```

The m_ButtonTestMessage represents the "Take Photo" button which the user presses to take a photo with the camera and have it sent back to the Android for display.

Button   m_ButtonTestMessage;

The m_TakePhotoButtonActive variable is true if the "Take Photo" button is active and able to be pressed. If it is false then the button is not active and can not be pressed.

boolean m_TakePhotoButtonActive = false;

The TakePhotoCallbackState enumeration describes the two stages of taking a photo. The first stage is getting the size of the captured image from the Arduino and the second stage is getting the actual image data from the Arduino using the image size to determine when all the binary data has been transmitted.

enum TakePhotoCallbackState

{

    GetImageSize,

    GetImageData

};

The m_TakePhotoState variable holds the current stage that the "Take Photo" command is in.

TakePhotoCallbackState m_TakePhotoState;

The m_ImageView represents the window on the Android user interface where photos transmitted from the Arduino will be displayed.

ImageView m_ImageView = null;

The m_TakePhotoCallbackDone variable is true if a Bluetooth command related to the "Take Photo" command has just completed and is in need of processing. It is false otherwise.

boolean m_TakePhotoCallbackDone = false;

The Resolution enumeration has the values of VGA, QVGA, QQVGA which refers to the resolution at which the user wants to captures images. These also form the commands that are transmitted to the Arduino.

enum Resolution

{

    VGA,

    QVGA,

    QQVGA

};

The AndroidStorage enumeration is StoreYes if storing received images is turned on. If the value is StoreNo then received images from the Arduino are not stored on the Android's file system.

enum AndroidStorage

```
{
    StoreYes,

    StoreNo
}
```

The m_AndroidStorage variable determines if incoming pictures from the Arduino are stored on the Android's file system. The default value is that pictures are not stored.

AndroidStorage m_AndroidStorage = AndroidStorage.StoreNo;

The m_PhotoData variable points to the received binary photo image data.

byte[] m_PhotoData;

The m_CameraDir holds the Android picture directory designated by the Android operating system.

File m_CameraDir;

The m_AndroidFileGroupID and m_AndroidFileItemID variables help identify which files are located on the Android file system.

int m_AndroidFileGroupID = 95;

int m_AndroidFileItemID = 35;

The m_AndroidFileNames string array holds a complete list of the names of the files located in the Android's picture directory.

String[] m_AndroidFileNames = new String[m_MaxFileNames];

The m_SurveillanceActive variable is true if camera surveillance is active and photos are continuously taken by the ArduCAM camera and sent to the Android.

boolean m_SurveillanceActive = false;

The m_SurveillanceFrameNumber variable holds the current number of the surveillance photo taken.

int m_SurveillanceFrameNumber = 0;

The m_SurveillanceFileName is the file name that a photo is saved under if Android storage is turned on.

String m_SurveillanceFileName = "";

The m_SurveillanceSFX variable is true if a sound effect is to be played every time a new photo is displayed on the Android when surveillance is active. The default is false which means no sound effect is played when surveillance is active.

boolean m_SurveillanceSFX = false;

The m_SendSystemStopCommand variable is true when a System Stop command needs to be sent to the Arduino and false otherwise.

boolean m_SendSystemStopCommand = false;

The onCreate() Function

The onCreate() function is called when the Android application first begins. We covered this function in Chapter 4 when discussing the Simple Intruder Alarm System. I have highlighted in bold print the additions needed when adding an ArduCAM camera to the system.

The onCreate() function requires the additions of :

1. Code to create and define a "Take Photo" button that commands the camera attached to the Arduino to take a photo and transmit it to the Android. More specifically:

    a. The button is found from the list Android user interfaces by looking it up by its resource id.

    b. The button is disabled or greyed out so that the user cannot activate it.

    c. The command to be sent to the Arduino is defined as the current resolution which is either VGA, QVGA, or QQVGA.

    d. A sound effect is played on the Android to indicate that the button has been pressed.

    e. The data structures that hold the incoming data from the Arduino are reset in the Bluetooth handler class object.

    f. Sets the command in the message handler to "GetImageSize". When a response from the Arduino is received it will be processed based on this command. That is, the size of the image that is captured will be processed.

    g. Set the stage of the "Take Photo" command to the "GetImageSize" stage.

    h. If there is a valid Bluetooth connection between the Android device and the Arduino then write the command from Step c. which is VGA, QVGA, or QQVGA to the Arduino.

    i. Disable the "Take Photo" button.

2. Code to initialize the Android photo storage system by calling InitExternalStorage().

3. Code to create and initialize the window on the Android user interface that displays the currently selected camera resolution that the user wishes future photos to be captured in.

See Listing 6-8.

Listing 6-8. The onCreate() function

@Override

protected void onCreate(Bundle savedInstanceState) {

```java
super.onCreate(savedInstanceState);

setContentView(R.layout.activity_main);


this.setRequestedOrientation(ActivityInfo.SCREEN_ORIENTATION_PORTRAIT);


// Initialize Text to Speech engine

TextToSpeechInit();


// Init Sounds

CreateSoundPool();

CreateSound(this);


// YUV Resource File to Bitmap Test

LoadYUVToImageViewDisplay(R.drawable.tulips_yuyv422_prog_packed_qcif); // Ok


// Initialize Test Message Button

m_ButtonTestMessage = (Button) findViewById(R.id.TestMessageButton);

m_ButtonTestMessage.setEnabled(false);

m_ButtonTestMessage.setOnClickListener(new View.OnClickListener()

{
        public void onClick(View v)

        {
                // Perform action on click

                m_DebugMsg += "Send Test Message Button Clicked!! \n";


                String Command = m_Resolution.toString();

                m_Alert1SFX.PlaySound(false);


                // Set Up Data Handler
```

```java
                            m_BluetoothMessageHandler.ResetData();
                            m_BluetoothMessageHandler.SetCommand("GetImageSize");
                            m_TakePhotoState = TakePhotoCallbackState.GetImageSize;


                            if (m_ClientConnectThread != null)
                            {
                                if (m_ClientConnectThread.GetConnectedThread() != null)
                                {
                                    m_TakePhotoButtonActive = false;
                                    m_ButtonTestMessage.setEnabled(false);
                                    m_ClientConnectThread.GetConnectedThread().write(Command.getBytes());
                                    m_DebugMsg += "Writing Take Photo Command Message to Arduino!! \n";
                                }
                                else
                                {
                                    m_DebugMsg += "ConnectedThread is Null!! \n";
                                }
                            }
                            else
                            {
                                m_DebugMsg += "ClientConnectThread is Null!! \n";
                            }
                            m_DebugMsgView.setText(m_DebugMsg.toCharArray(), 0, m_DebugMsg.length());
                        }
                });

        // Initialize the Output Message Window
        m_OutputMsgView = (EditText) findViewById(R.id.outputmsg);
```

```java
        m_OutputMsg = "BlueTooth Devices: \n";
        m_OutputMsgView.setText(m_OutputMsg.toCharArray(), 0, m_OutputMsg.length());
        m_OutputMsgView.setFocusable(false);
        m_OutputMsgView.setOnClickListener(new View.OnClickListener()
        {
                public void onClick(View v) {
                        m_Alert2SFX.StopSound();
                }
        });


        // Initialize the Debug Message Window
        m_DebugMsgView = (EditText) findViewById(R.id.debugmsg);
        m_DebugMsg = "Debug Output: \n";
        m_DebugMsgView.setText(m_DebugMsg.toCharArray(), 0, m_DebugMsg.length());
        m_DebugMsgView.setFocusable(false);
        m_DebugMsgView.setOnClickListener(new View.OnClickListener()
        {
                public void onClick(View v) {
                        m_Alert2SFX.StopSound();
                }
        });


        // Init External Storage
        InitExternalStorage();


        // Initialize the Information Text window
        m_InfoTextView = (TextView) findViewById(R.id.textView1);
        UpdateCommandTextView();
        m_InfoTextView.setFocusable(false);
```

```
// Initializes the Android to Arduino Message Handling
InitializeMessageHandling();


// Test for Bluetooth Adapter
// If Bluetooth is supported and it is not enabled then ask the user
// if he wants to turn it on.
m_BluetoothAdapter = BluetoothAdapter.getDefaultAdapter();
if (m_BluetoothAdapter == null) {
    // Device does not support Bluetooth
    Log.e("BlueToothTest", "Device does not support Bluetooth!");

    m_DebugMsg += "Device does not support Bluetooth! \n";
    m_DebugMsgView.setText(m_DebugMsg.toCharArray(), 0, m_DebugMsg.length());
}
else
{
    Log.d("BlueToothTest","Device DOES Support Bluetooth!!!!");
    m_DebugMsg += "Device DOES Support Bluetooth!!!! \n";
    m_DebugMsgView.setText(m_DebugMsg.toCharArray(), 0, m_DebugMsg.length());


    // Test if Bluetooth is enabled
    if (!m_BluetoothAdapter.isEnabled())
    {
        Log.d("BlueToothTest","Blue Tooth is NOT Enabled. Request to activate BlueTooth!!!!!!!");
        m_DebugMsg += "Blue Tooth is NOT Enabled. Request to activate BlueTooth !!!!!!! \n";
        m_DebugMsgView.setText(m_DebugMsg.toCharArray(), 0, m_DebugMsg.length());


        Intent enableBtIntent = new Intent(BluetoothAdapter.ACTION_REQUEST_ENABLE);
```

```
                    startActivityForResult(enableBtIntent, REQUEST_ENABLE_BT);
            }
            else
            {
                    // Blue Tooth is Enabled so initialize it.
                    InitializeBlueTooth();
            }
        }
}
```

The InitExternalStorage() function retrieves the directory that the images from the ArduCAM will be stored in.

The function does the following:

1. Tests if the storage is writable by calling the isExternalStorageWritable() function. Display a message in the Debug Window with the result.

2. The final path to the directory is retrieved by calling the GetAlbumStorageDir(Environment.DIRECTORY_PICTURES, "ArduinoCameraPics") function with the input parameters for the directory type which is the pictures directory and the album name which will be "ArduinoCameraPics".

3. The path is also printed to the Debug Window.

See Listing 6-9.

Listing 6-9. The InitExternalStorage() function

```
void InitExternalStorage()
{
        // Test for External Storage and check if writable
        if (isExternalStorageWritable())
        {
                String Message = "External Storage is PRESENT and WRITABLE ...\n";
                AddDebugMessage(Message);
        }
        else
```

```
            {
                    String Message = "External Storage is GONE or NOT Writable !!!! \n";

                    AddDebugMessage(Message);

            }

            // Get/Create Directory for Arduino Camera pictures

            m_CameraDir = GetAlbumStorageDir(Environment.DIRECTORY_PICTURES, "ArduinoCameraPics");

            Path = "Arduino Camera Pics Directory: " + m_CameraDir.getAbsolutePath() + " ...\n";

            AddDebugMessage(Path);

}
```

The isExternalStorageWritable() function determines if the external storage on the Android file system is writable by:

1. Getting the external storage state of the Android system by calling the Environment.getExternalStorageState() function.

2. If the result is equal to "MEDIA_MOUNTED" then we can save the images to storage.

See Listing 6-10.

Listing 6-10. The isExternalStorageWritable() function

```
public boolean isExternalStorageWritable()

{
            String state = Environment.getExternalStorageState();

            if (Environment.MEDIA_MOUNTED.equals(state))

            {
                return true;

            }

            return false;

}
```

The GetAlbumStorageDir() function creates and returns a new File object by:

1. Creating a new File class object using the input parameters of DirectoryType and AlbumName. This File object retrieves the system file path for the DirectoryType such as Pictures that is on the Android file system and adds on the AlbumName to the end of the path.

2. Attempting to create this new path by calling the file.mkdirs() function which makes all the directories in the path contained in the file variable from Step 1. Add a message to the Debug Window if the path could not be created.

See Listing 6-11.

Listing 6-11. The GetAlbumStorageDir() function

```java
public File GetAlbumStorageDir(String DirectoryType, String AlbumName)
{
    // Get the directory for the user's album

    File file = new File(Environment.getExternalStoragePublicDirectory(DirectoryType), AlbumName);

    if (!file.mkdirs()) {

        Log.e("BLUETOOTHTEST", "External Storage Album Directory not created !!!!");

        String Error = "External Storage Album Directory not created !!!!\n";

        AddDebugMessage(Error);

    }

    return file;

}
```

The UpdateCommandTextView() function updates the information in the window that displays the camera resolution by converting the m_Resolution variable that holds an enumeration to a string and then updating the window on the Android user interface that displays this information. Note that there are extra unused parameters shown that are not currently used for the version of the camera we are going to use in this book which is the ArduCAM Mini digital camera. See Listing 6-12.

Listing 6-12. The UpdateCommandTextView() function

```java
void UpdateCommandTextView()

{

    // Get Current Camera Resolution and set text view

    String Info = m_Resolution.toString() + "\n" + m_FPS.toString() + "\n" + m_AWB.toString() + "\n" + m_AEC.toString() + "\n" + m_YUVMatrix.toString() + "\n" + m_Denoise.toString() + "\n" + m_EdgeEnhancement.toString() + "\n" + m_ABLC.toString();

    int length = Info.length();

    m_InfoTextView.setText(Info.toCharArray(), 0, length);
```

}

The onOptionsItemSelected() Function

The onOptionsItemSelected() function serves to process the user's menu selections. This section discusses the additions to this function that were needed because of the addition of the ArduCAM Mini digital camera and the new functions associated with it in our ArduCAM intruder alarm system.

The additions to this function are:

1. Loading in an Android Image file that the user has clicked on by calling the LoadInAndroidImage(item) function with the menu item object representing the image filename that the user selected.

2. If the user has selected VGA as the resolution for taking photos, then change the resolution of the photo to take to VGA and update the window that displays the camera parameters.

3. If the user has selected QVGA as the resolution for taking photos, then change the resolution of the photo to take to QVGA and update the window that displays the camera parameters.

4. If the user has selected QQVGA as the resolution for taking photos, then change the resolution of the photo to take to QQVGA and update the window that displays the camera parameters.

5. If the user selects the menu item to turn on image storage then set m_AndroidStorage = AndroidStorage.StoreYes.

6. If the user selects the menu item to turn off the image storage then set m_AndroidStorage = AndroidStorage.StoreNo.

7. If the user selects the "Load JPEG Image From Android" menu then the function ProcessLoadJpgImagesAndroid(item) is called to populate the menu with the JPEG files found in our "ArduinoCameraPics" directory that we created earlier to store the pictures from the ArduCAM camera.

8. If the user turns on the surveillance then the ProcessSurveillanceSelection(true) function is called with the true parameter.

9. If the user turns off the surveillance then the ProcessSurveillanceSelection(false) function is called with the false parameter.

10. If the user turns on the surveillance sound effects on then m_SurveillanceSFX is set to true.

11. If the user turns off the surveillance sound effects on then m_SurveillanceSFX is set to false.

See Listing 6-13.

Listing 6-13. The onOptionsItemSelected() function

@Override

```java
public boolean onOptionsItemSelected(MenuItem item) {
    // Process file selection located on Android
    if (item.getGroupId() == 95)
    {
        LoadInAndroidImage(item);
        return true;
    }

    // Handle item selection
    switch (item.getItemId()) {
        case R.id.SetVGA:
            ChangeResolution(Resolution.VGA);
            UpdateCommandTextView();
            return true;

        case R.id.SetQVGA:
            ChangeResolution(Resolution.QVGA);
            UpdateCommandTextView();
            return true;

        case R.id.SetQQVGA:
            ChangeResolution(Resolution.QQVGA);
            UpdateCommandTextView();
            return true;

        case R.id.androidstoreyes:
            m_AndroidStorage = AndroidStorage.StoreYes;
            m_DebugMsg = "";
            String Yes = "Images will now be saved to Android Storage ...\n";
```

```
                AddDebugMessage(Yes);
        return true;


    case R.id.androidstoreno:
                m_AndroidStorage = AndroidStorage.StoreNo;
                m_DebugMsg = "";
                String No = "Images will NOT be saved to Android Storage ...\n";
                AddDebugMessage(No);
        return true;


    case R.id.loadandroidimage:
                ProcessLoadJpgImagesAndroid(item);
        return true;


    case R.id.surveillanceon:
                ProcessSurveillanceSelection(true);
        return true;


    case R.id.surveillanceoff:
                ProcessSurveillanceSelection(false);
        return true;


    case R.id.surveillancesfxon:
                m_SurveillanceSFX = true;
        return true;


    case R.id.surveillancesfxoff:
                m_SurveillanceSFX = false;
        return true;
```

```java
            // System
            case R.id.activatesystem:
                    SendSystemArmCommmandToArduino(true);
                return true;


            case R.id.deactivatesystem:
                    SendSystemArmCommmandToArduino(false);
                return true;


            case R.id.getsystemstatus:
                    SendGetStatusCommandToArduino();
                return true;


            case R.id.getintruderstatus:
                    SendGetIntruderStatusCommandToArduino();
                return true;


            // Call Out
            case R.id.activatecallout:
                    SetCallOutStatus(true);
                return true;


            case R.id.deactivatecallout:
                    SetCallOutStatus(false);
                return true;


            default:
                return super.onOptionsItemSelected(item);
```

				}

}

The LoadInAndroidImage() function loads in the file selected by the user and displays the image on the Android user interface.

The function does the following:

1. Continues processing the selected menu item only if it is located on the Android file system as indicated by the item's group id being equal to the Android file group ID.

2. Gets the filename from the menu item by converting the menu item's title into a string.

3. Loads the image data from the file into the m_PhotoData variable by calling the LoadImageDataAndroid(m_CameraDir, filename) function with the parameters for the directory and the filename of the file in the directory to load.

4. Finally, the file image data is converted into a viewable format and displayed on the Android device by calling the LoadJpegPhotoUI(m_PhotoData) function with the JPEG picture data.

See Listing 6-14.

Listing 6-14. The LoadInAndroidImage() function

```
void LoadInAndroidImage(MenuItem item)
{
        String filename = "";


        if (item.getGroupId() == m_AndroidFileGroupID)
        {
                filename = item.getTitle().toString();

                Log.e("Main Activity", "Android image File Selected = " + "'" + filename + "'");

                LoadImageDataAndroid(m_CameraDir, filename);

                LoadJpegPhotoUI(m_PhotoData);
        }
}
```

The LoadImageDataAndroid() loads in the image data from the file by:

1. Creating a new File object using the directory the file is in and the filename.

2. Creating a buffered input stream using the File object from step 1 and reading in all the file data and storing it in the m_PhotoData array variable.

See Listing 6-15.

Listing 6-15. The LoadImageDataAndroid() function

```java
void LoadImageDataAndroid(File Dir, String filename)
{
    // Loads binary image data from the filename located in Dir directory from Android
    // and puts it into the m_PhotoData array used to hold the current image data being
    // displayed on Android

    // Open File to write out image data
    File file = new File(Dir, filename);

    if (file != null)
    {
        Log.e("MainActivity", "Android Image File path constructed ....");
        Log.e("MainActivity", "Directory = " + Dir);
        Log.e("MainActivity", "Filename = " + filename);
    }

    InputStream iStream = null;

    int InputByte = 0;
    int Index = 0;

    try
    {
        iStream = new BufferedInputStream(new FileInputStream(file));
```

```
                    while (InputByte >= 0)

                    {

                                InputByte = iStream.read();

                                if (InputByte != -1)

                                {

                                            m_PhotoData[Index] = (byte)InputByte;

                                            Index++;

                                }

                    }

        }

        catch (IOException e)

        {

                    Log.e("MainActivity", "Cannot load image data from Android...");

        }

}
```

The LoadJpegPhotoUI() function displays a JPEG image on the Android device's user interface based on the input JpegData.

The function does the following:

1. The data in the JpegData byte array is decoded into a bitmap image. This is done by calling the BitmapFactory.decodeByteArray(JpegData, offset, length) function with the image data, the offset from the beginning of the array to where the image data starts and the length of the data that forms the image.

2. The ImageView window that will hold the image is found by calling the findViewById(R.id.imageView1) function with the resource id of the window we want to put the image in.

3. The image is put into the window by calling the ImageView.setImageBitmap(BM) function with the bitmap image generated from the JPEG data from Step 1.

4. If the image from Step 1 is null which means that the decoding failed then issue an error in the log window by calling the Log.e() function.

See Listing 6-16.

Listing 6-16. The LoadJpegPhotoUI() function

```
// For use with ArduCAM
void LoadJpegPhotoUI(byte[] JpegData)
{
    /*
        public static Bitmap decodeByteArray (byte[] data, int offset, int length)
        Added in API level 1

        Decode an immutable bitmap from the specified byte array.

        Parameters:
            data    byte array of compressed image data
            offset  offset into imageData for where the decoder should begin parsing.
            length  the number of bytes, beginning at offset, to parse

        Returns:
        The decoded bitmap, or null if the image could not be decoded.
    */

    Bitmap BM = null;
    int offset = 0;
    int length = JpegData.length;

    BM = BitmapFactory.decodeByteArray(JpegData, offset, length);
    ImageView ImageView = (ImageView) findViewById(R.id.imageView1);
    if (BM != null)
    {
        ImageView.setImageBitmap(BM);
    }
```

```
            else
            {
                    Log.e("MainActivity", "ImageView Bitmap Creation Failed!!!");
            }
    }
```

The ChangeResolution() function sets the global m_Resolution variable to the input resolution. There is some other code that changes other properties such as the width of the photo but they are not used with the ArduCAM camera and have been left here as a guide on how to change these parameters if you need to for another camera. See Listing 6-17.

Listing 6-17. The ChangeResolution() function

```
void ChangeResolution(Resolution Res)
{
        switch (Res)
        {
                case VGA:
                        m_PhotoWidth    = 640;
                        m_PhotoHeight   = 480;
                        m_BytesPerPixel = 2;
                break;


                case QVGA:
                        m_PhotoWidth    = 320;
                        m_PhotoHeight   = 240;
                        m_BytesPerPixel = 2;
                break;


                case QQVGA:
                        m_PhotoWidth    = 160;
                        m_PhotoHeight   = 120;
```

```
                    m_BytesPerPixel = 2;

            break;

    }


        m_PhotoSize = (m_PhotoWidth * m_PhotoHeight) * m_BytesPerPixel;

    m_Resolution = Res;

}
```

The ProcessLoadJpgImagesAndroid() function loads in the filenames of the JPEG files located in the directory we specifically created for the camera images for our intruder alarm system.

The function does the following:

1. Gets the submenu from the input menuitem input parameter by assigning to the variable submenu1 the result from the menuitem.getSubMenu() function call.

2. Retrieves a list of filenames in the directory we created for storing camera images by calling the m_CameraDir.list() function.

3. Assign the number of files in the directory to the m_NumberFileNames variable.

4. Clear all the entries in the "Load JPEG Image From Android" menu that lists the JPEG files in the camera image storage directory we previously defined by calling the submenu1.clear() function.

5. For each filename in the m_AndroidFileNames String array test to see if the filename's extension indicates that it is a JPEG file by calling the IsFileJPEG(m_AndroidFileNames[i]) function on each element in the array.

6. If the filename has a .jpg extension then add this filename to the "Load JPEG Image From Android" sub menu by calling the submenu1.add(m_AndroidFileGroupID, m_AndroidFileItemID, i, m_AndroidFileNames[i]) function with input parameters that identify this entry as located on the Android file system.

See Listing 6-18.

Listing 6-18. The ProcessLoadJpgImagesAndroid() function

```
void ProcessLoadJpgImagesAndroid(MenuItem menuitem)

{

        Log.e("MainActivity","In ProcessLoadJpgImagesAndroid");

        SubMenu submenu1 = menuitem.getSubMenu();
```

```java
            // Load Android JPG directory
            m_AndroidFileNames = m_CameraDir.list();
            m_NumberFileNames = m_AndroidFileNames.length;


            if (submenu1 != null)
            {
                    // Add files from Arduino SD Card Here.
                    submenu1.clear();
                    for (int i = 0; i < m_NumberFileNames; i++)
                    {
                            if (IsFileJPEG(m_AndroidFileNames[i]))
                            {
                                    submenu1.add(m_AndroidFileGroupID, m_AndroidFileItemID, i, m_AndroidFileNames[i]);
                            }
                    }
            }
            else
            {
                    Log.e("MainActivity", "ERROR, Can not get SubMenu");
            }
}
```

The IsFileJPEG() function returns true if the input filename extension is ".jpg" and false otherwise.

The function does the following:

1. Find the starting position of the filename extension by calling the filename.indexOf('.') function. The filename variable is the input string parameter.

2. Get the substring of the filename starting at one position after the position of the "." which will be the filename extension.

3. Convert this substring to lower case.

4. Trim this substring of any trailing whitespace characters such as spaces, new line characters, or control characters.

5. Compare this substring to "jpg". If there is a match return true to indicate that a JPEG file has been found. Otherwise, return false.

See Listing 6-19.

Listing 6-19. The IsFileJPEG() function

```
boolean IsFileJPEG(String filename)
{
    boolean ValidExtension = false;
    String Extension = filename;
    String ExtensionProcessed = "";
    String ExtensionTrimmed = "";

    // File extension is valid if .yuv
    int indexdot = filename.indexOf('.');
    int start = indexdot + 1;
    if (indexdot > 0)
    {
        ExtensionProcessed = Extension.substring(start);
        String LowerCaseExtensionProcessed = ExtensionProcessed.toLowerCase();
        ExtensionTrimmed = LowerCaseExtensionProcessed.trim();
        int result = ExtensionTrimmed.compareTo("jpg");
        if (result == 0)
        {
            ValidExtension = true;
        }
    }
    return ValidExtension;
}
```

The ProcessSurveillanceSelection() function activates or deactivates the surveillance system based on the input parameter.

The function does the following:

1. If surveillance is already on and the user is trying to turn on surveillance then do nothing since surveillance is already on and exit the function.

2. If the user selects to turn surveillance on then activate the surveillance, reset the surveillance frame number and execute the SendTakePhotoCommand() function in order to perform a "Take Photo" command to start off the camera surveillance on the Arduino side.

3. If the user selects to turn surveillance off then the surveillance is turned off.

See Listing 6-20.

Listing 6-20. The ProcessSurveillanceSelection function

```
void ProcessSurveillanceSelection(boolean Activated)
{
    if (m_SurveillanceActive && Activated)
    {
        // surveillance is already activated so do nothing
        return;
    }

    // If surveillance set to active
    if (Activated)
    {
        // Surveillance is set to on.
        m_SurveillanceActive = true;

        // Disable "Take Photo" button while video surveillance is active
        m_TakePhotoButtonActive = false;

        // Reset Surveillance Frame Number
```

```
                m_SurveillanceFrameNumber = 0;


                // Send TakePhoto Command to Arduino

                SendTakePhotoCommand();

        }

        else

        {

                m_SurveillanceActive = false;

        }
}
```

The SendSystemArmCommmandToArduino() function sends a System Start or System Stop command to the Arduino. I discussed this function in the previous chapter on the Simple Intruder Alarm System. I have highlighted in bold the additions to this function.

The additions to this function are:

1. If the input parameter is false which means that a System Stop command is to be issued and surveillance is currently active then we need to wait until the current image frame is transmitted to the Android before sending a System Stop command to the Arduino.

2. If there is no surveillance currently running that means there is no image data being transmitted to the Android and thus we can send the System Stop command right now.

3. If we are sending a System Start command then we need to disable the "Take Photo" button so that the user can not take a photo and interfere with the listening for an intruder alert message from the Arduino.

4. If we are sending a System Stop command then we can enable the "Take Photo" button.

See Listing 6-21.

Listing 6-21. The SendSystemArmCommmandToArduino() function

```
void SendSystemArmCommmandToArduino(boolean Active)

{

        String Command = "";


        // Send Command To Arduino to Arm/Disarm system

        if (Active)
```

```
            {
                    m_TTS.speak("Arming System", TextToSpeech.QUEUE_ADD, null);
                    Command = "SystemStart";
            }
            else
            {
                    if (m_SurveillanceActive)
                    {
                            // Flag to send System Stop command after current surveillance cycle is finished.
                            m_SendSystemStopCommand = true;
                            return;
                    }
                    else
                    {
                            // If surveillance is not active then send systemstop command now
                            m_TTS.speak("Disarming System", TextToSpeech.QUEUE_ADD, null);
                            Command = "SystemStop";
                    }
            }

            // Send Command to Arduino
            m_DebugMsg += "Sending System command to Arduino!! \n";

            // Set Up Data Handler
            m_BluetoothMessageHandler.SetCommand(Command);
            if (m_ClientConnectThread != null)
            {
                    if (m_ClientConnectThread.GetConnectedThread() != null)
                    {
```

```
                    if (Active)
                    {
                            m_ButtonTestMessage.setEnabled(false);
                            m_TakePhotoButtonActive = false;
                    }
                    else
                    {
                            m_ButtonTestMessage.setEnabled(true);
                            m_TakePhotoButtonActive = true;
                    }


                    m_ClientConnectThread.GetConnectedThread().write(Command.getBytes());
                    m_DebugMsg += "Writing System Start/Stop command to Arduino!! \n";
            }
            else
            {
                    m_DebugMsg += "ConnectedThread is Null!! \n";
            }
    }
    else
    {
            m_DebugMsg += "ClientConnectThread is Null!! \n";
    }
    m_DebugMsgView.setText(m_DebugMsg.toCharArray(), 0, m_DebugMsg.length());
}
```

The Android Menu

For the ArduCAM version of the intruder system we needed to add some more menu items such as:

1. Resolution Settings – Sets the camera resolution for taking pictures.

a. Set VGA Camera Mode – Sets the VGA resolution 640 by 480 pixels.

   b. Set QVGA Camera Mode – Sets the QVGA resolution 320 by 240 pixels

   c. Set the QQVGA Camera Mode – Sets the QQVGA resolution of 160 by 120 pixels.

2. Android Storage Settings:

   a. Save Images to Android Storage – Saves incoming images from the Arduino on the Android file system.

   b. Do Not Save Images to Android Storage – Turns the saving of incoming images from the Arduino to the Android file system off. Default is not to save any incoming images.

3. Load JPEG Image from Android – Displays a list of JPEG images saved from our ArduCAM camera if any.

4. Surveillance Settings:

   a. Activate Surveillance – Turns on the surveillance for our system and continually transmits images from our camera attached to the Arduino to our Android device.

   b. Deactivate Surveillance – Turns off the surveillance.

   c. Turn Surveillance Sound Effects ON – Plays a sound effect after each new image frame is finished transmitting to Android from Arduino.

   d. Turn Surveillance Sound Effects OFF – Turns off the playing of a sound effect after each new image frame is transmitted.

See Listing 6-22.

Listing 6-22. XML code for the Android Menu

```xml
<menu xmlns:android="http://schemas.android.com/apk/res/android" >

   <item

      android:id="@+id/camera_settings"

      android:title="Resolution Settings" >

      <menu>

         <item android:id="@+id/SetVGA" android:title="Set VGA Camera Mode" /><item android:id="@+id/SetQVGA" android:title="Set QVGA Camera Mode" />

         <item android:id="@+id/SetQQVGA" android:title="Set QQVGA Camera Mode" />

      </menu>

   </item>
```

```xml
<item android:id="@+id/androidstoragesettings" android:title="Android Storage Settings">
    <menu>
        <item android:id="@+id/androidstoreyes" android:title="Save Images to Android Storage"/>
        <item android:id="@+id/androidstoreno" android:title="Do NOT Save Images to Android Storage"/>
    </menu>
</item>


<item android:id="@+id/loadandroidimage" android:title="Load JPEG Image From Android">
    <menu></menu>
</item>


    <item android:id="@+id/surveillance" android:title="Surveillance Settings">
        <menu>
            <item android:id="@+id/surveillanceon" android:title="Activate Surveillance"/>
            <item android:id="@+id/surveillanceoff" android:title="Deactivate Surveillance"/>
            <item android:id="@+id/surveillancesfxon" android:title="Turn Surveillance Sound Effects ON"/>
            <item android:id="@+id/surveillancesfxoff" android:title="Turn Surveillance Sound Effects OFF"/>
        </menu>
    </item>


    <item android:id="@+id/systemsettings" android:title="System Settings">
        <menu>
            <item android:id="@+id/activatesystem" android:title="Arm System"/>
            <item android:id="@+id/deactivatesystem" android:title="Unarm System"/>
            <item android:id="@+id/getsystemstatus" android:title="Get System Status"/>
            <item android:id="@+id/getintruderstatus" android:title="Get Intruder Status"/>
        </menu>
    </item>
```

```xml
        <item android:id="@+id/calloutsettings" android:title="Call Out Settings">
          <menu>
            <item android:id="@+id/activatecallout" android:title="Activate Emergency Call Out"/>
            <item android:id="@+id/deactivatecallout" android:title="Deactivate Emergency Call Out"/>
          </menu>
        </item>
</menu>
```

The Android Graphic User Interface (GUI)

The additions needed for our ArduCAM camera intruder alarm are:

1. The imageView1 ImageView class window that displays the photo being transmitted from the Arduino.

2. The TestMessageButton Button class button that is the "Take Photo" button

3. The textView1 TextEdit class window that displays the resolution of the next image to take and as well as other information that is not currently used but serves as placeholders for future expansion.

See Listing 6-23.

Listing 6-23. The Android Graphic User Interface XML

```xml
<RelativeLayout xmlns:android="http://schemas.android.com/apk/res/android"
    xmlns:tools="http://schemas.android.com/tools"
    android:layout_width="match_parent"
    android:layout_height="match_parent"
    tools:context=".MainActivity" >

    <ImageView
        android:id="@+id/imageView1"
        android:layout_width="320dp"
        android:layout_height="240dp"
```

```xml
        android:layout_alignParentLeft="true"
        android:layout_alignParentTop="true"
        android:src="@drawable/ic_launcher" />

    <Button
        android:id="@+id/TestMessageButton"
        android:layout_width="wrap_content"
        android:layout_height="wrap_content"
        android:layout_alignParentRight="true"
        android:layout_alignTop="@+id/outputmsg"
        android:text="@string/TestButtonText" />

    <EditText
        android:id="@+id/outputmsg"
        android:layout_width="wrap_content"
        android:layout_height="wrap_content"
        android:layout_alignParentLeft="true"
        android:layout_below="@+id/imageView1"
        android:layout_toLeftOf="@+id/TestMessageButton"
        android:ems="10"
        android:lines="3"
        android:maxLines="3"
        android:text="@string/hello_world" />

    <EditText
        android:id="@+id/debugmsg"
        android:layout_width="wrap_content"
        android:layout_height="wrap_content"
        android:layout_alignParentLeft="true"
```

```
        android:layout_alignRight="@+id/outputmsg"

        android:layout_below="@+id/outputmsg"

        android:ems="10"

        android:lines="4"

        android:maxLines="4"

        android:text="@string/Debug_Message" />


    <EditText

        android:id="@+id/textView1"

        android:layout_width="wrap_content"

        android:layout_height="wrap_content"

        android:layout_alignParentRight="true"

        android:layout_alignTop="@+id/debugmsg"

        android:layout_toRightOf="@+id/debugmsg"

        android:ems="10"

        android:scrollHorizontally="true"

        android:scrollbarStyle="outsideOverlay"

        android:scrollbars="vertical"

        android:text="TextView"

        android:textSize="11dp" />


</RelativeLayout>
```

The BluetoothMessageHandler Class

The BluetoothMessageHandler class needs to be changed in order to use the ArduCAM camera. I previously discussed this class in the chapter on the Simple Intruder Alarm System.

The ReceiveMessage() function which receives Bluetooth data from the Arduino needs to be modified with the following additions in order to use the camera:

1. If the command that is expecting a response is the GetImageSize command then call the ProcessGetImageSizeCommand(NumberBytes, Message) function to process the response from the Arduino.

2. If the command that is expecting a response is the GetImageData command then call the ProcessGetImageDataCommand(NumberBytes, Message) function to process the response from the Arduino.

See Listing 6-24.

Listing 6-24. The ReceiveMessage function

```
void ReceiveMessage(int NumberBytes, byte[] Message)
{
        // Process Incoming Data
        // Assume incoming data is associated with the current m_Command variable
        // and process the Message accordingly.
        if (m_Command == "GetImageSize")
        {
                ProcessGetImageSizeCommand(NumberBytes, Message);
        }
        else
        if (m_Command == "GetImageData")
        {
                ProcessGetImageDataCommand(NumberBytes, Message);
        }
        else
        if (m_Command == "SystemStart")
        {
                // System is monitoring for Intruder Alerts.
                ProcessSystemStartCommand(NumberBytes,Message);
        }
        else
        if (m_Command == "SystemStop")
        {
                ProcessSystemStopCommand(NumberBytes,Message);
```

```
            }
            else
        if (m_Command == "GetSystemStatus")
        {
                ProcessGetStatusCommand(NumberBytes,Message);
        }
        else
        if (m_Command == "GetIntruderStatus")
        {
                ProcessGetIntruderStatusCommand(NumberBytes, Message);
        }
        else
        if (m_Command == "GetPhoneNumber")
        {
                ProcessGetPhoneNumberCommand(NumberBytes, Message);
        }
        else
        {
                Log.e("BlueToothTest" , "Error - Command for Data Receive Not Found!!!!!!");
        }
}
```

The ProcessGetImageSizeCommand() function processes the GetImageSize command by:

1. Receiving the incoming text data by calling the ReceiveTextData(NumberBytes, Message) function.

2. If all of the text data has been read in then:

    a. Call the m_MainActivity.TakePhotoCommandCallback() function to notify the MainActivity class object that there is incoming data from the Take Photo command that needs to be processed.

    b. Call the m_MainActivity.runOnUiThread(m_MainActivity) function to execute the processing of the incoming data.

See Listing 6-25.

Listing 6-25. The ProcessGetImageSizeCommand() function

```
void ProcessGetImageSizeCommand(int NumberBytes, byte[] Message)
{
        boolean FinishedReceivingText = ReceiveTextData(NumberBytes, Message);
        if (FinishedReceivingText)
        {
                // Process Remote Directory Command
                m_MainActivity.TakePhotoCommandCallback();
                Log.e("BlueTooth Test","TakePhotoCommandCallback() called for GetImageSize Command... ");


                // Update the User Interface
                m_MainActivity.runOnUiThread(m_MainActivity);
        }
}
```

The TakePhotoCommandCallback() function is called from the BluetoothMessageHandler class and does the following:

1. Sets up the incoming data relating to the Take Photo command for processing by setting m_TakePhotoCallbackDone to true.

2. If the Take Photo command is in the GetImageData stage then retrieve a pointer to the binary image data and assign it to m_PhotoData.

See Listing 6-26.

Listing 6-26. The MainActivity class TakePhotoCommandCallback() function

```
void TakePhotoCommandCallback()
{
        Log.e("MainActivity", "In TakePhotoCommandCallback()");


        // Photo is now ready for viewing
        m_TakePhotoCallbackDone = true;
```

```
        if (m_TakePhotoState == TakePhotoCallbackState.GetImageData)

        {

                m_PhotoData = m_BluetoothMessageHandler.GetBinaryData();

        }

}
```

The ProcessGetImageDataCommand() function retrieves and processes the data associated with the GetImageData command by:

1. Calling the ReceiveBinaryData(NumberBytes, Message) function to read in the binary image data from the Arduino.

2. If the data has been fully transmitted:

    a. Call the m_MainActivity.TakePhotoCommandCallback() function to notify the MainActivity class that there is new data to process for the Take Photo command.

    b. Call the m_MainActivity.runOnUiThread(m_MainActivity) function to execute the processing of the new data.

See Listing 6-27.

Listing 6-27. The ProcessGetImageDataCommand() function

```
void ProcessGetImageDataCommand(int NumberBytes, byte[] Message)

{

        boolean FinishedReceivingData = ReceiveBinaryData(NumberBytes, Message);

        if (FinishedReceivingData)

        {

                // Process Take Picture Command

                m_MainActivity.TakePhotoCommandCallback();

                Log.e("BlueTooth Test","TakePhotoCommandCallback called for GetImageData ... ");


                // Update the User Interface

                m_MainActivity.runOnUiThread(m_MainActivity);

        }
```

}

The MainActivity's run() Function

The run() function in the MainActivity class is called when the m_MainActivity.runOnUiThread(m_MainActivity) function is called from the BluetoothMessageHandler class. This function is used here to process incoming data from the Arduino and to update the Android device's user interface. I covered this function previously and in this section I will cover the additions needed to use the camera.

The additions to the run() function are:

1. If new data related to the Take Photo command needs to be processed and the current state of the command is expecting to receive the size of the image then call the ProcessGetImageSizeCallback() function. After returning from the function advance the state of the command to expect the data from the actual image to be returned.

2. If new data related to the Take Photo command needs to be processed and the current state of the command is expecting the image data then:

    1. The LoadJpegPhotoUI(m_PhotoData) function is called to convert the image data into a viewable image and display it on the Android device's user interface.

    2. If Android storage for incoming photos is turned on then call the SaveArudinoPicJPEG() function to save the photo to the Android file system.

    3. If surveillance is active then:

        1. Disable the Take Photo button

        2. If playing sound effects for surveillance is turned on then play a sound effect.

        3. The surveillance frame number is updated and printed to the Debug Window.

        4. Display the surveillance filename to the Debug Window if Android storage is turned on.

        5. If the user has requested a System Stop command be issued then turn surveillance off and call the SendSystemArmCommmandToArduino(false) function with the false parameter to send a System Stop command to the Arduino.

        6. If the user has not requested a System Stop command then continue with surveillance by calling the SendTakePhotoCommand() function to take another photo with the camera.

    4. If surveillance is not active then enable the Take Photo button and play a continuous sound effect to notify the user that the image he requested is now being displayed.

3. If the Take Photo button should be active then enable it otherwise disable it.

See Listing 6-28.

Listing 6-28. The run() function

```
// Updates data in User Interface
public void run()
{
        Log.e("MainActivity", "..........In RUN() Callback FUNCTION...............");


        // Take Photo Button callback
        if(m_TakePhotoCallbackDone)
        {
                if (m_TakePhotoState == TakePhotoCallbackState.GetImageSize)
                {
                        Log.e("MainActivity", "Procesing Image Size ...");
                        ProcessGetImageSizeCallback();
                        m_TakePhotoState = TakePhotoCallbackState.GetImageData;
                }
                else
                if (m_TakePhotoState == TakePhotoCallbackState.GetImageData)
                {
                        // Display New Photo received from the Arduino after pressing the
                        // Take Photo Button.
                        Log.e("MainActivity", "Loading JPEG data into UI");
                        LoadJpegPhotoUI(m_PhotoData);

                        // Save the Arduino Pic to Android External File System.
                        m_SurveillanceFileName = "";
                        if (m_AndroidStorage == AndroidStorage.StoreYes)
                        {
                                SaveArudinoPicJPEG();
                        }
```

```
                        // If surveillance is active
                        if (m_SurveillanceActive)
                        {
                                m_ButtonTestMessage.setEnabled(false);
                                if (m_SurveillanceSFX)
                                {
                                        m_Alert1SFX.PlaySound(false);
                                }
                                m_SurveillanceFrameNumber++;
                                m_DebugMsg = "";
                                m_DebugMsg += "Surveillance Frame Number: " + m_SurveillanceFrameNumber + "\n";

                                if (m_AndroidStorage == AndroidStorage.StoreYes)
                                {
                                        m_DebugMsg += "Surveillance File Name: " + m_SurveillanceFileName + "\n";
                                }

                                m_DebugMsgView.setText(m_DebugMsg.toCharArray(), 0, m_DebugMsg.length());

                                // Check to see if user has unarmed alarm system
                                if (m_SendSystemStopCommand)
                                {
                                        // Turn Surveillance off
                                        m_SurveillanceActive = false;

                                        // Send SystemStop Command
                                        SendSystemArmCommmandToArduino(false);
```

```
                              // Finished sending system stop command
                              m_SendSystemStopCommand = false;
                        }
                        else
                        {
                              // Send another take photo command to Arduino
                              SendTakePhotoCommand();
                        }
                  }
                  else
                  {
                        m_ButtonTestMessage.setEnabled(true);
                        m_Alert2SFX.PlaySound(true);
                  }
            }
            m_TakePhotoCallbackDone = false;
      }


      // SystemStart Command Callback
      if (m_StartSystemCommandFinished)
      {
            ProcessStartSystemCommandResult();
            m_StartSystemCommandFinished = false;
      }


      // SystemStop Command Callback.
      if (m_StopSystemCommandFinished)
      {
```

```
            ProcessStopSystemCommandResult();

            m_StopSystemCommandFinished = false;

    }


    // Get Phone Number From Arduino

    if (m_SendGetPhoneNumberCommand)

    {

            SendGetPhoneNumberMessage();

            m_SendGetPhoneNumberCommand = false;

    }


    // GetPhoneNumber Command Callback

    if (m_GetPhoneNumberCommandFinished)

    {

            ProcessGetPhoneNumberCommand();

            m_GetPhoneNumberCommandFinished = false;

    }


    // GetSystemStatus Command Callback

    if (m_GetSystemStatusCommandFinished)

    {

            // System has returned the status of the Alarm system on the

            // Arduino side

            ProcessGetSystemStatusValue();

            m_GetSystemStatusCommandFinished = false;

    }


    // Get Intruder Status Command Callback

    if (m_GetIntruderStatusCommandFinished)
```

```
                {
                        ProcessGetIntruderStatusValue();
                        m_GetIntruderStatusCommandFinished = false;
                }


                // Set Take Photo Button active state
                if (m_TakePhotoButtonActive == true)
                {
                        m_ButtonTestMessage.setEnabled(true);
                }
                else
                {
                        m_ButtonTestMessage.setEnabled(false);
                }
}
```

The ProcessGetImageSizeCallback() function does the following:

1. Retrieves the incoming text data from the BluetoothMessageHandler class object.

2. Trims the text data to remove the newline character, spaces, and other whitespace characters.

3. Converts the String data into a number using the Integer.parseInt(DataTrimmed) function with the value from Step 2 as an input parameter and assigns the result to m_PhotoSize.

4. Calls the SendGetImageDataCommand() to send the command to the Arduino to transmit the actual binary image data.

See Listing 6-29.

Listing 6-29. The ProcessGetImageSizeCallback() function

```
void ProcessGetImageSizeCallback()
{
        String Data = m_BluetoothMessageHandler.GetStringData();
        String DataTrimmed = Data.trim();
```

```
            m_PhotoSize = Integer.parseInt(DataTrimmed);

            SendGetImageDataCommand();
}
```

The SendGetImageDataCommand() function sends a GetImageData command to the Arduino by doing the following:

1. Resets the data in the BluetoothMessageHandler object.

2. Sets the length of the binary data that the Android will receive by calling the m_BluetoothMessageHandler.SetDataReceiveLength(m_PhotoSize) function with the size of the photo expected.

3. The command in the BluetoothMessageHandler is set to GetImageData.

4. If the Bluetooth connection between the Android device and the Arduino is valid then disable the Take Photo button and transmit the command to the Arduino.

See Listing 6-30.

Listing 6-30. The SendGetImageDataCommand() function

```
void SendGetImageDataCommand()
{
        String Command = "GetImageData";

        // Set Up Data Handler
        m_BluetoothMessageHandler.ResetData();
        m_BluetoothMessageHandler.SetDataReceiveLength(m_PhotoSize);
        Log.e("MainActivity", "Photosize = " + m_PhotoSize);
        m_BluetoothMessageHandler.SetCommand(Command);
        if (m_ClientConnectThread != null)
        {
                if (m_ClientConnectThread.GetConnectedThread() != null)
                {
                        m_ButtonTestMessage.setEnabled(false);
                        m_ClientConnectThread.GetConnectedThread().write(Command.getBytes());
```

```
                    m_DebugMsg += "Writing GetImageData command to Arduino!! \n";
                }
                else
                {
                    m_DebugMsg += "ConnectedThread is Null!! \n";
                }
            }
            else
            {
                m_DebugMsg += "ClientConnectThread is Null!! \n";
            }
            m_DebugMsgView.setText(m_DebugMsg.toCharArray(), 0, m_DebugMsg.length());
}
```

The SaveArudinoPicJPEG() function saves the photo that was transmitted to the Android from the Arduino by:

1. Creating the beginning of the filename from the image resolution such as VGA, QVGA or QQVGA.

2. Creating the date by first getting a Calendar class object by calling the Calendar.getInstance() function. Then using this object to get the DayOfYear, Hour, Minute, and Second and adding this to the end of the filename.

3. Adding the ".jpg" extension to the filename.

4. Creating a new file using the filename in the directory specified by m_CameraDir.

5. Writing out the image data that is in the m_PhotoData array to the file.

6. If surveillance is active then assign the filename created to the m_SurveillanceFileName String variable.

See Listing 6-31.

Listing 6-31. The SaveArudinoPicJPEG() function

```
void SaveArudinoPicJPEG()
{
        // Main Filename
```

```java
String Filename = m_Resolution.toString() + "_";


// Add Date
Calendar rightNow = Calendar.getInstance();
int DayOfYear = rightNow.get(Calendar.DAY_OF_YEAR);
int Hour = rightNow.get(Calendar.HOUR_OF_DAY);
int Minute = rightNow.get(Calendar.MINUTE);
int Second = rightNow.get(Calendar.SECOND);
String Date = DayOfYear + "_" + Hour + "_" + Minute + "_" + Second;


// Add JPG Extension
String Extension = ".jpg";


// Build Final Filename
Filename += Date;
Filename += Extension;


// Open File to write out image data
File file = new File(m_CameraDir, Filename);


try
{
        OutputStream os = new FileOutputStream(file);
        os.write(m_PhotoData, 0, m_PhotoSize);
        os.close();
}
catch (IOException e)
{
        // Unable to create file, likely because external storage is
```

```
                // not currently mounted.

                Log.e("ExternalStorage", "Error writing " + file, e);
        }


        // If surveillance is active then record the filename that the photo was saved to.
        if (m_SurveillanceActive)
        {
                m_SurveillanceFileName = Filename;
        }
}
```

The SendTakePhotoCommand() function starts a Take Photo command by:

1. Setting the command to send to the Arduino to the current user selected resolution such as VGA, QVGA, or QQVGA.

2. Resets the data in the message handler by calling the m_BluetoothMessageHandler.ResetData() function.

3. Sets the command in the message handler to one that matches the type of return value expected from the Arduino which is the GetImageSize command. The data type expected is the size of the photo in bytes that was just taken by the camera.

4. The state of the Take Photo command is set to the GetImageSize state.

5. The Take Photo button is disabled.

6. The command created in Step 1 is converted from a String object into an array of bytes and transmitted to the Arduino.

See Listing 6-32.

Listing 6-32. The SendTakePhotoCommand() function

```
void SendTakePhotoCommand()
{
        String Command = m_Resolution.toString();


        // Set Up Data Handler
        m_BluetoothMessageHandler.ResetData();
```

```
                m_BluetoothMessageHandler.SetCommand("GetImageSize");

                m_TakePhotoState = TakePhotoCallbackState.GetImageSize;

                if (m_ClientConnectThread != null)

                {

                        if (m_ClientConnectThread.GetConnectedThread() != null)

                        {

                                m_ButtonTestMessage.setEnabled(false);

                                m_ClientConnectThread.GetConnectedThread().write(Command.getBytes());

                                m_DebugMsg += "Writing Take Photo Command Message to Arduino!! \n";

                        }
                        else

                        {

                                m_DebugMsg += "ConnectedThread is Null!! \n";

                        }

                }
                else

                {

                        m_DebugMsg += "ClientConnectThread is Null!! \n";

                }

                m_DebugMsgView.setText(m_DebugMsg.toCharArray(), 0, m_DebugMsg.length());

}
```

The ProcessStartSystemCommandResult() function processes the responses from the Arduino to the Android device's SystemStart command that turns on the intruder alarm system. I discussed this function previously for the simple alarm system in Chapter 4.

In order to use this function with our camera the following additions were made:

1. Turns on the surveillance and actually begins the video surveillance by calling the SendTakePhotoCommand() function.

See Listing 6-33.

Listing 6-33. The ProcessStartSystemCommandResult() function

```
void ProcessStartSystemCommandResult()
{
        String Data = m_BluetoothMessageHandler.GetStringData();

        String DataTrimmed = Data.trim();

        if (DataTrimmed.equalsIgnoreCase("SystemStartOK"))

        {
                m_ArduinoAlarmSystemON = true;

                m_TTS.speak("System Has Started on Arduino Side", TextToSpeech.QUEUE_ADD, null);

                m_DebugMsg = "";

                AddDebugMessage("System Has Stared on Arduino Side");
        }

        else

        if (DataTrimmed.equalsIgnoreCase("IR_MOTION_ALERT"))

        {
                m_IR_MOTION_ALERT = true;

                m_DebugMsg = "";

                AddDebugMessage("INTRUDER DETECTED!!!!");

                if (m_CallOutActive)

                {
                        CallEmergencyPhoneNumber();
                }

                else

                {
                        // Repeat message

                        for (int i = 0; i < m_EmergencyMessageRepeatTimes ; i++)

                        {
                                m_TTS.speak("Intruder Alert Intruder has been detected by the motion sensor", TextToSpeech.QUEUE_ADD, null);
```

```
                            }

                    m_Alert2SFX.PlaySound(true);

                }


            // Activate the Alarm System's Surveillance Function

            // Set surveillance status to active and send a take photo command to the Arduino to start the surveillance

            m_SurveillanceActive = true;

            SendTakePhotoCommand();

        }

        else

        {

            m_TTS.speak("ERROR in System Start command response from Arduino", TextToSpeech.QUEUE_ADD, null);

            AddDebugMessage("ERROR in System Start command response from Arduino!!!!");

        }

    }
```

Android Manifest

The Android Manifest XML file needs to be modified.

The WRITE_EXTERNAL_STORAGE permission needs to be added in order for the Android device to be able to store incoming images onto the Android file system.

```
<uses-permission android:name="android.permission.WRITE_EXTERNAL_STORAGE" />
```

See Listing 6-34.

Listing 6-34. Android Manifest XML

```xml
<?xml version="1.0" encoding="utf-8"?>

<manifest xmlns:android="http://schemas.android.com/apk/res/android"

    package="com.example.bluetoothtest"

    android:versionCode="1"

    android:versionName="1.0" >
```

```xml
<uses-sdk
    android:minSdkVersion="8"
    android:targetSdkVersion="16" />

<application
    android:allowBackup="true"
    android:icon="@drawable/ic_launcher"
    android:label="@string/app_name"
    android:theme="@style/AppTheme" >
    <activity
        android:name="com.example.bluetoothtest.MainActivity"
        android:label="@string/app_name" >
        <intent-filter>
            <action android:name="android.intent.action.MAIN" />

            <category android:name="android.intent.category.LAUNCHER" />
        </intent-filter>
    </activity>
</application>

<uses-permission android:name="android.permission.CALL_PHONE" />
<uses-permission android:name="android.permission.BLUETOOTH" />
<uses-permission android:name="android.permission.BLUETOOTH_ADMIN" />
<uses-permission android:name="android.permission.WRITE_EXTERNAL_STORAGE" />

</manifest>
```

Summary

In this chapter I covered the ArduCAM based intruder alarm system. I started by giving an overview of the type of ArduCAM camera that was used for this project and then showed you how to install the needed software libraries. Next, I covered the additional C/C++ Arduino code needed to support the ArduCAM camera on the Arduino side. Finally, I discussed the additional Java Android code needed to support the ArduCAM camera on the Android side.

Chapter 7

Hands on Example: Building an ArduCAM Intruder Alarm / Surveillance System

In this chapter I show you how to build an intruder/surveillance alarm system that uses the ArduCAM digital camera. I start off with an overview of the ArduCAM Mini digital camera we will use for this project. Next, I discuss the ArduCAM software development library and its installation. Then, I present a wiring diagram of the intruder alarm/surveillance system that you will be building followed by detailed instructions on how exactly to connect everything together. Next, I tell you where to find the official web site for this book as well as where to download the related executables and source code. Finally, I present a step by step quick start guide to using your new wireless ArduCAM intruder and surveillance alarm system.

ArduCAM Mini Camera Hardware Overview

The ArduCAM Mini 2 Mega Pixel ov2640 digital camera that we will use for this project is made by ArduCAM and the official web site is located at:

http://www.arducam.com

The ArduCAM Mini is an easy to use camera that uses the standard SPI and I2C interfaces for the Arduino. The pins on the ArduCAM Mini are as follows:

- CS – "Chip Select" or "Slave Select" which is the pin on each device that the master can use to enable and disable specific devices. When a device's Slave Select pin is low, it communicates with the master. When it's high, it ignores the master. This allows you to have multiple SPI devices sharing the same MISO, MOSI, and CLK lines.

- MOSI – "Master Out Slave In" which is the master line for sending data to the peripherals.

- MISO – "Master In Slave Out" which is the slave line for sending data to the master

- SCK – "Serial Clock" which is the clock pulses which synchronize data transmission generated by the master

- GND – Ground pin.

- VCC – Input Voltage pin that accepts either 3.3 volts or 5.0 volts.

- SDA – 'Data Line" for the I2C interface.

- SCL – "Clock Line" for the I2C interface

See Figure 6-1.

Figure 6-1. ArduCAM Mini Module Camera Shield with ov2640 2 Mega Pixels Lens

The ArduCAM Mini 2 Mega Pixels ov2640 camera can be purchased in many places including on Amazon.

The listing in Amazon is "Arducam Mini Module Camera Shield with OV2640 2 Megapixels Lens for Arduino UNO Mega2560 Board" and the cost is around $30.

ArduCAM Mini Camera Library Software Installation

In order to use the ArduCAM Mini Camera with your Arduino you will need to download and install the ArduCAM libraries. You will need to go to the ArduCAM web site listed in the previous section and download the library.

Once you download the zip file you will need to uncompress it using a program like 7-zip and install the two directories "ArduCAM" and "UTFT4ArduCAM_SPI" under the "libraries" directory for Arduino. For example, on my Windows XP system I have installed the ArduCAM libraries in my "Program Files/Arduino/libraries" directory by copying the two directories to this "libraries" directory. After doing this you should be able to compile source code that includes the ArduCAM library.

ArduCAM Intruder Alarm System Overview

The general set up of our ArduCAM intruder alarm/surveillance system consists of an Android device communicating with an Arduino equipped with a Bluetooth Adapter, an infrared motion detector, and an ArduCAM Mini Module ov2640 Camera Shield with 2 Mega Pixel lens. The Android sends commands to the Arduino and the Arduino responds to these commands with text data or binary data. When an intruder is detected a text message will be sent to the Android device. The Android device will then either call out to an emergency phone number or issue an audible alert so that the home owner can be notified that the alarm system has detected an intruder. Surveillance will then automatically start and photos will be taken by the camera and sent to the Android until the alarm system is deactivated. See Figure 7-1.

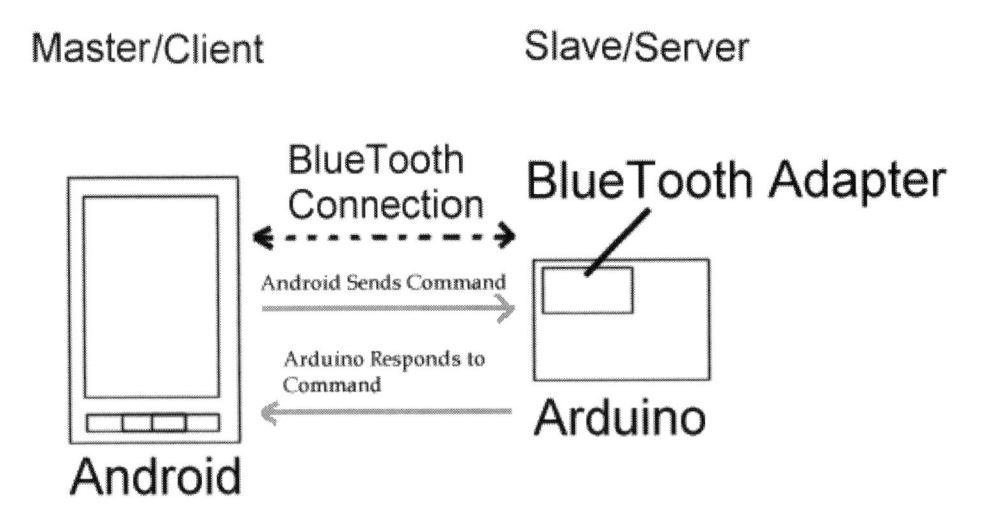

Figure 7-1. System overview

Wiring Diagram

In order to create the alarm system in this chapter you will need to connect the components of the system in this example which are:

- An HC-SR501 Pyroelectric Infrared (PIR) Motion Sensor Detector for Arduino
- An HC-06 4 pin Serial Bluetooth Adapter Module for Arduino
- An Arduino UNO
- An ArduCAM Mini Module Camera Shield with OV2640 2 Mega Pixels Lens

- Breadboard
- Male to Male and Male to Female Jumper wires

See Figure 7-2.

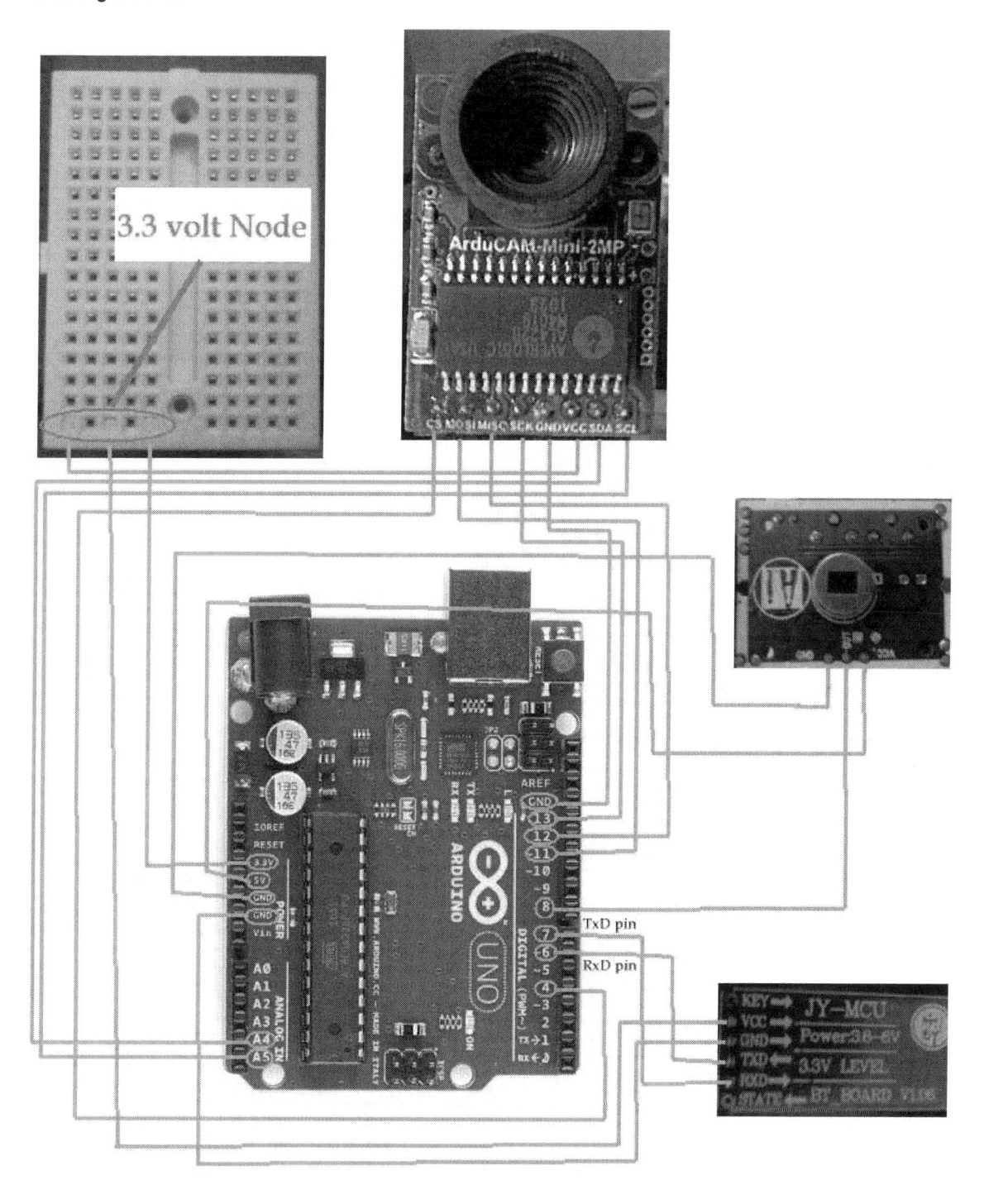

Figure 7-2. Wiring Diagram

All of these parts are available on Amazon.com now as I am typing these words. Ebay is also a good place to look for these parts.

Detailed Connection Instructions

To connect the motion sensor to the Arduino UNO you need to:

- Connect the GND pin on the sensor to a GND pin on the Arduino.
- Connect the OUT pin on the sensor to pin 8 on the Arduino.
- Connect the VCC pin on the sensor to the 5 volt pin on the Arduino

To connect the Bluetooth adapter to the Arduino UNO you need to:

- Connect the VCC pin on the sensor to the 3.3 volt node on the breadboard.
- Connect the GND pin on the sensor to the GND pin on the Arduino.
- Connect the TXD pin on the sensor to pin 6 on the Arduino.
- Connect the RXD pin on the sensor to pin 7 on the Arduino.

To connect the ArduCAM Mini camera to the Arduino UNO you need to:

- Connect the VCC pin on the camera to the 3.3 volt breadboard node
- Connect the GND pin on the camera to one of the GND pins on the Arduino.
- Connect the SDA (data line) on the camera to the analog input pin 4 on the Arduino.
- Connect the SCL (clock line) on the camera to the analog input pin 5 on the Arduino.
- Connect the MOSI pin on the camera to pin 11 on the Arduino.
- Connect the MISO pin on the camera to pin 12 on the Arduino.
- Connect the SCK pin on the camera to pin 13 on the Arduino.
- Connect the CS pin on the camera to pin 4 on the Arduino.

Official Support Web Site

The official support web site for this book can be found at:

http://www.psycho-sphere.com/diy.html

Downloading the Final Android APK installable

The final Android executable which is an .apk file can be downloaded directly at:

http://www.psycho-sphere.com/ArduCAMBurglarAlarmV10.apk

Copy this file onto your Android device and run it to begin the installation. Usually your Android device has a file manager where you can begin installing an Android .apk file by clicking on it. Also, you may need to allow installation from unknown sources or non-android market sources to install this program to your device. This option is under the Android's system settings.

After installation you should see a new icon representing this program on your Android device.

Downloading the Code

The Arduino Uno source code for this alarm system is located at:

http://www.psycho-sphere.com/ArduCAM_Mini_2MP_IntruderAlarmSystemUNO.zip

The Android source code for this alarm system is located at:

http://www.psycho-sphere.com/WorkSpaceArduCAMIntruderAlarmSystemV10.zip

Importing Android Code

If you want to install the program from the source code then you will first need to unzip the file using a program such as 7 zip (which is free) and then import the code.

If you are using the ADT bundle you will need to select File->Import->Android->Existing Android Code Into Workspace to get started importing the code. Follow the directions in the subsequent window dialogs to import the code into your current workspace.

After you import the code you can then select "MainActivity" in the leftmost window pane in the Integrated Development Environment or IDE and select Run->Run from the menu to begin installation of the program. In the next window popup select "Android Application" to run the program as an Android application.

If you are using Android Studio you will have to first convert the old Eclipse ADT project code files into the new form. Please refer to the latest information at:

http://developer.android.com/sdk/installing/migrate.html

Quick Start User's Guide

1. Connect the ArduCAM camera, Arduino UNO, motion detector, and Bluetooth adapter as shown in Figure 7-2.

2. Make sure you have the Arduino Integrated Development Environment installed on your computer system.

3. Download and unzip the Arduino source code using a program such as 7-zip (which is free at http://www.7-zip.org).

4. Open the Arduino IDE and load in the source code for the alarm system.

5. Change the emergency phone number in the Arduino source code to the number you want your Android cell phone to call out to when the intruder alarm system detects an intruder.

    String PhoneNumber = "9876543210"; ← Change this number to your own emergency number

6. Save your customized program.

7. Compile and upload this program to your Arduino UNO that you have connected to your computer using your USB cable. See Chapter 1 for more information on setting up the Arduino and compiling and uploading Arduino programs.

8. Next, install the Android application for the alarm system by either downloading the .apk file and installing it or downloading the source code, importing it into your Android development system and then installing it to your Android device.

9. Start up the Serial Monitor on the Arduino IDE which brings up the Serial Monitor window where debug messages from the Arduino program are printed out.

10. You should see the following be displayed in the Serial Monitor window. Please note that the Bluetooth adapter should return a value of "OK" upon successful initialization.

Note: The code assumes that the Bluetooth Adapter is set for a speed of 9600 baud. If the speed is different from this value then you will need to change the Arduino code. See Chapter 3 for more information.

ArduCAM Android/Arduino Remote Intruder Alarm Wireless Camera System Version 1.0
Initializing ...

Software Serial RX, TX Pins are: 6,7

ArduCAM Starting ....

OV2640 detected ....

Initializing Bluetooth ...

OK

AT Command Test ... Success

Initializing IR Proximity Sensor ....

==

ArduCAM Android/Arduino Remote Intruder Alarm Wireless Camera System Version 1.0

This code to accompany the book entitled:

Home Security Systems DIY using Android and Arduino By Robert Chin

Please refer to the book for detailed explainations

of how to use this code, how to use the

ArduCAM camera as well as how to set up the camera with

the Arduino

==

11. Next, start up the Android application. If your Bluetooth is not currently activated then the application will attempt to start it up. You will see a popup window requesting permission to turn on Bluetooth. Click the "Yes" button to start up bluetooth. Also, make sure "Airplane Mode" is off before you try to turn on Bluetooth.

234

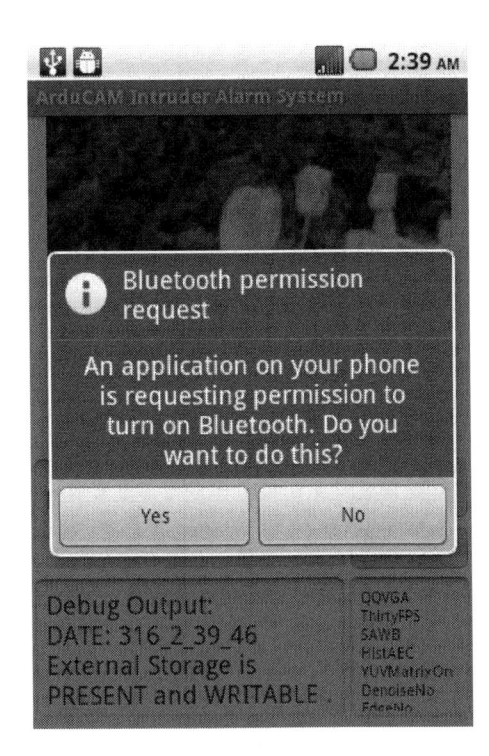

12. After clicking "Yes" you should see a message that the Android device is trying to turn on Bluetooth.

13. Next, the application will start up discovery for new Bluetooth enabled devices. When it finds the Arduino's Bluetooth adapter it will issue a pair request if the devices have not been previously paired.

14. Click the text field and enter the password which is generally "1234" and then click "Done" and then "OK". You may need to click on the "SYM" key first to activate the numerical keys.

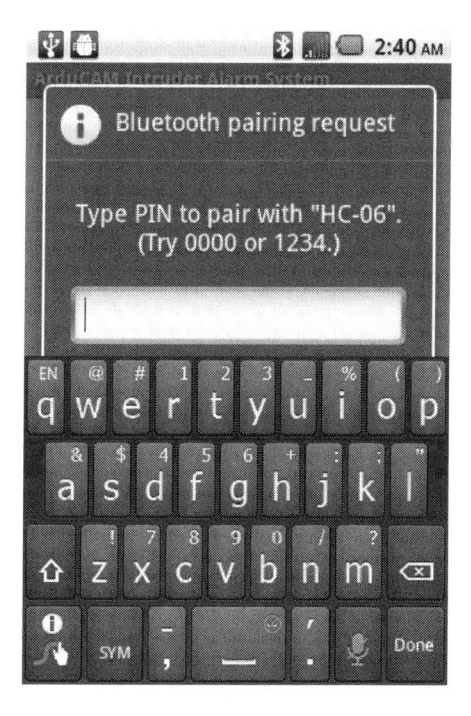

15. A message will be displayed saying that the devices are now paired and now a Bluetooth connection is being established between the Android device and the Arduino. After the connection is made you should hear vocal announcements as well as text updates in the top Output Window and the bottom Debug Window.

16. On the Arduino side you should see the debug output showing that the Android is requesting and then receiving the emergency phone number from the Arduino.

Raw Command from Android: GetPhoneNumber

Command is GetPhoneNumber ...

Command: GetPhoneNumber

Command GetPhoneNumber has been executed ...

17. The Arduino and Android devices are now connected. The default picture resolution is QQVGA. Let's take a picture by pressing the "Take Photo" button. The Take Photo command is broken down into two parts. The first part sends the command and then receives the image length in bytes of the photo that was captured. The second part involves transmitting the actual binary image data from the Arduino to the Android device. On the Arduino's Serial Monitor you should see something like the following.

Raw Command from Android: QQVGA

Command: QQVGA

.... Starting Image Capture ....

Start Capture

Capture Done!

FIFO RAW IMAGE SIZE = 3072

Transmitting Image Length = 3071

Command QQVGA Executed ...

Raw Command from Android: GetImageData

Command: GetImageData

READ_FIFO_LENGTH() = 3072

ByteCount = 3071

Total time used:3 seconds ...

Command GetImageData Executed ...

18. The photo should then be displayed on your Android device. See below.

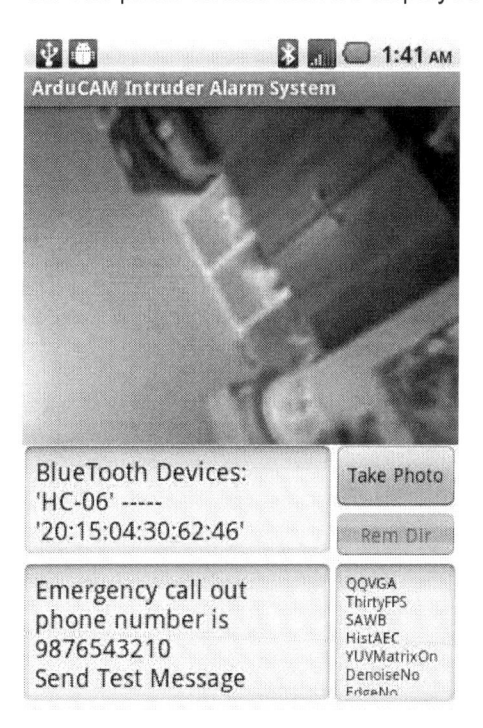

19. Press the Menu button on your Android device to bring up the menu system. Select the "Resolution" menu button to bring up the list of camera resolutions.

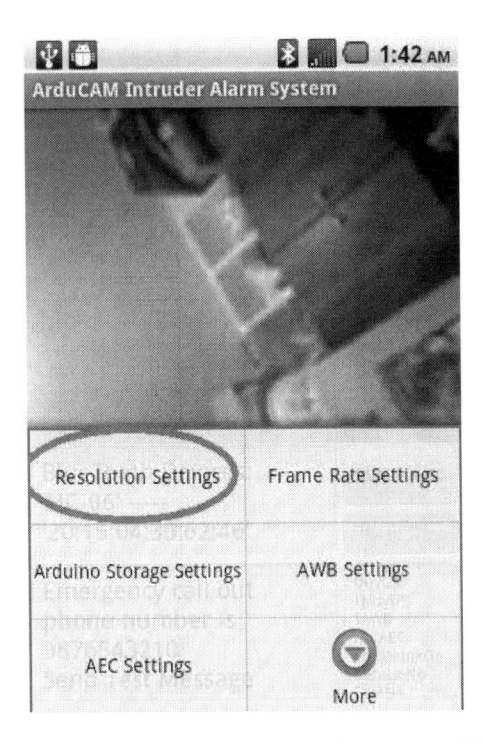

20. The list of camera resolutions are VGA (640 by 480 pixels), QVGA (320 by 240 pixels), and QQVGA (160 by 120 pixels).

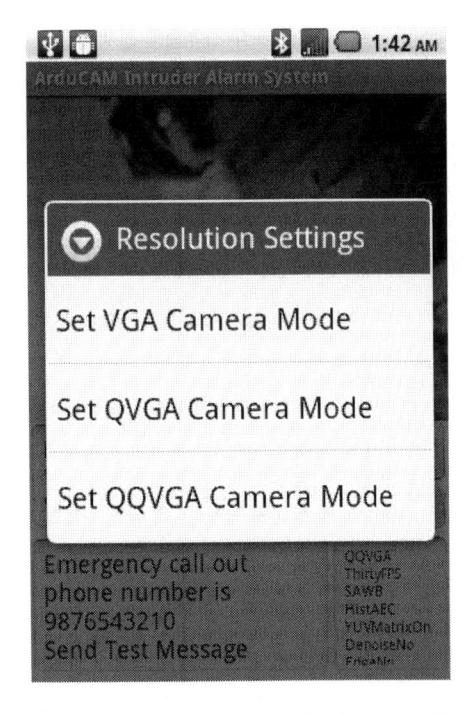

21. Select the "Set QVGA Camera Mode" selection. Press the "Take Photo" button. The Serial Monitor from the Arduino should display the following. The Serial Monitor should show the command which is QVGA, the image capture information, the captured image length, and the transfer information for the actual binary data for the image.

Raw Command from Android: QVGA

Command: QVGA

.... Starting Image Capture ....

Start Capture

Capture Done!

FIFO RAW IMAGE SIZE = 9216

Transmitting Image Length = 9215

Command QVGA Executed ...

Raw Command from Android: GetImageData

Command: GetImageData

READ_FIFO_LENGTH() = 9216

ByteCount = 9215

Total time used:9 seconds ...

Command GetImageData Executed ...

22. When the transfer is finished the QVGA resolution image should appear on your Android device. You should see a higher resolution image such as the one below.

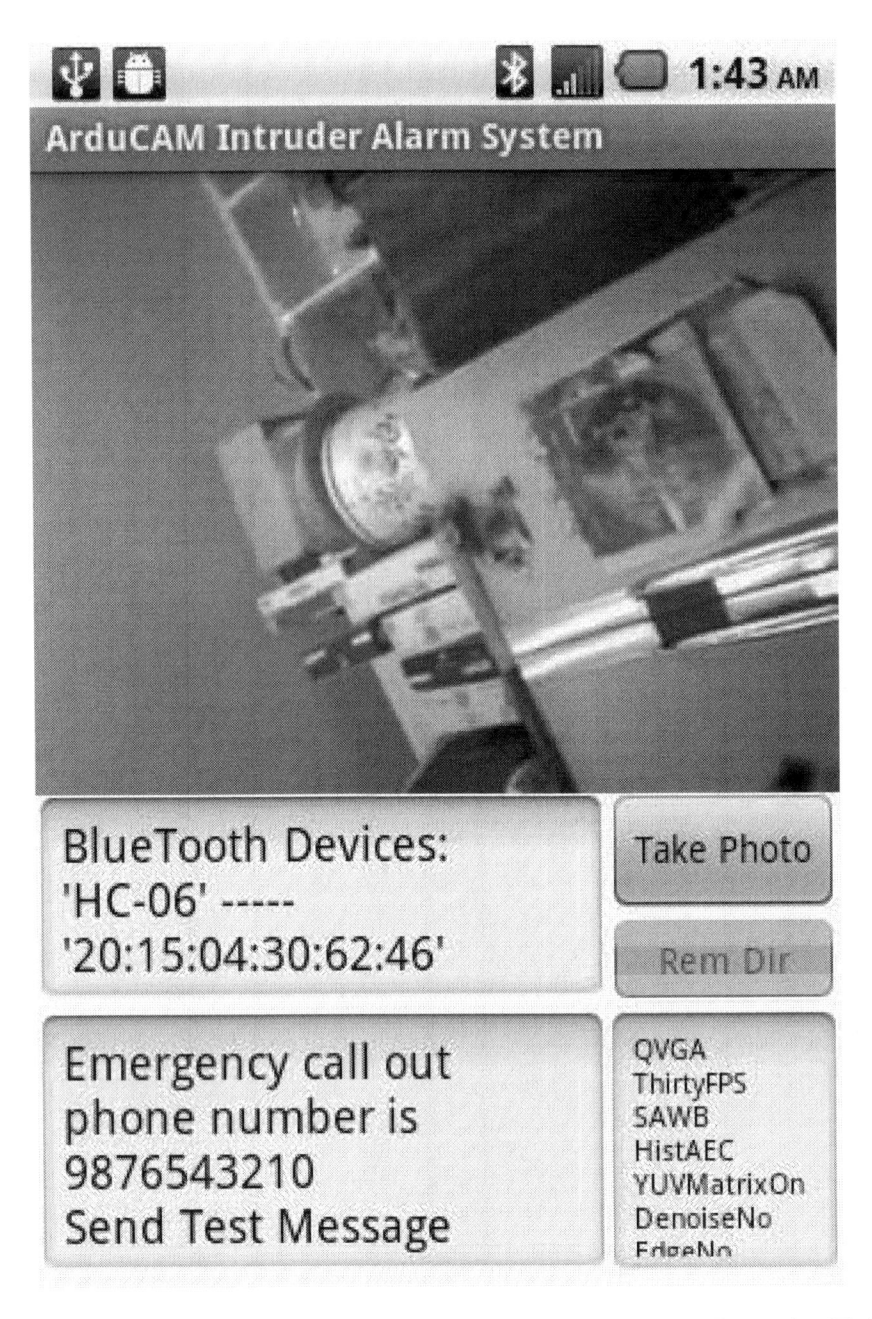

23. Press the menu button and select the VGA resolution. Press the "Take Photo" button again. Look at your Arduino Serial Monitor. You should see the VGA command, the image capture process, the GetImageData command, and the transfer of binary image data to the Android device.

Raw Command from Android: VGA

Command: VGA

.... Starting Image Capture ....

Start Capture

Capture Done!

FIFO RAW IMAGE SIZE = 19456

Transmitting Image Length = 19455

Command VGA Executed ...

Raw Command from Android: GetImageData

Command: GetImageData

READ_FIFO_LENGTH() = 19456

ByteCount = 19455

Total time used:20 seconds ...

Command GetImageData Executed ...

24. The VGA photo should be displayed on the Android device after the data transmission is finished. Notice the image appears more detailed than the QVGA photo which appears more detailed than the QQVGA photo. Also note that the transmission time is also longer at 20 seconds for VGA, 9 seconds for QVGA, and 3 seconds for QQVGA.

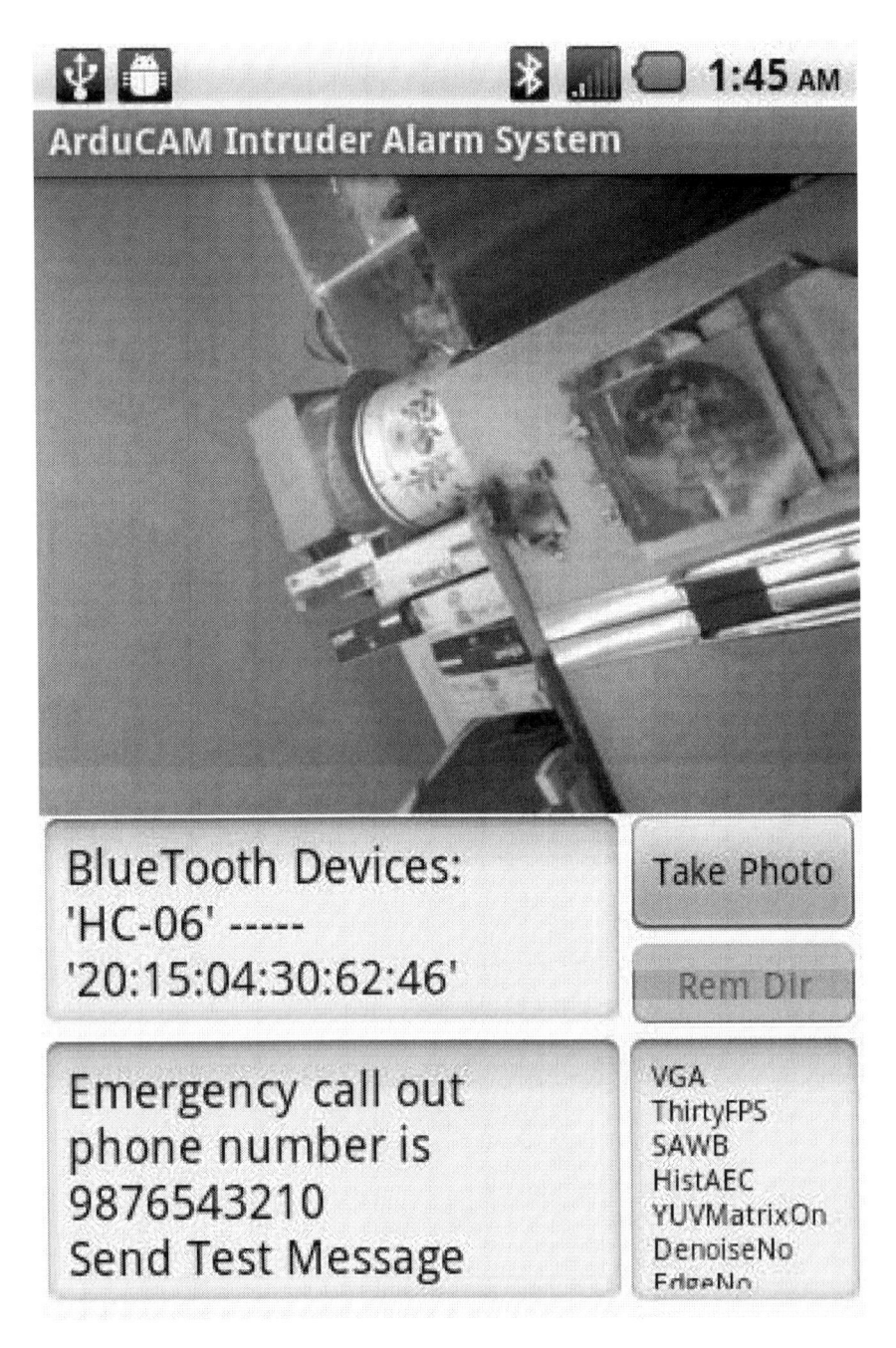

25. Set the resolution back to QQVGA and go to the main menu and select the "Surveillance Settings" menu.

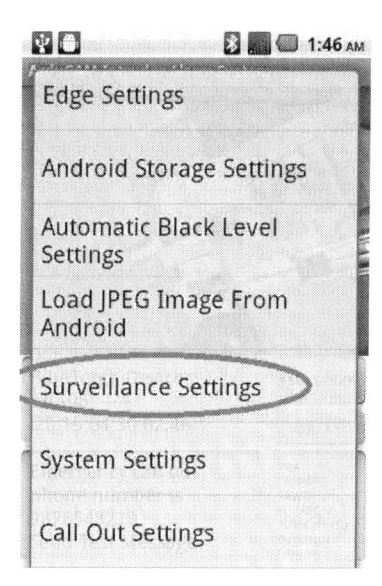

26. The Surveillance Settings menu should come up. Select the "Activate Surveillance" setting to activate surveillance. This starts the continuous taking of photos and the display of these photos on the Android device.

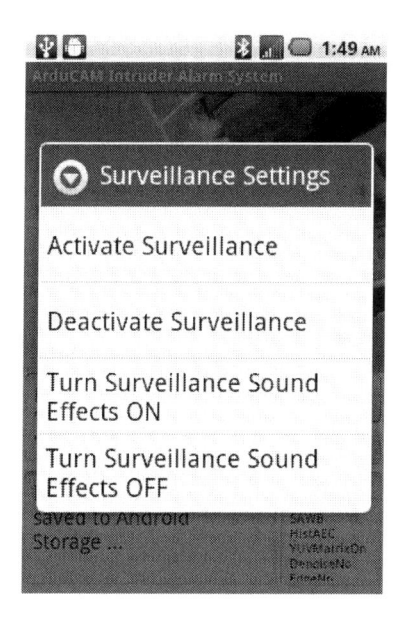

27. On the Serial Monitor for the Arduino you should see repeated photos being taken and transmitted to the Android device. The output should be similar to the following which is the same as pressing the "Take Photo" button multiple times. Instead of having to do this manually the program automatically takes photos continuously until the user turns surveillance off.

Raw Command from Android: QQVGA

Command: QQVGA

.... Starting Image Capture ....

Start Capture

Capture Done!

FIFO RAW IMAGE SIZE = 2048

Transmitting Image Length = 2047

Command QQVGA Executed ...

Raw Command from Android: GetImageData

Command: GetImageData

READ_FIFO_LENGTH() = 2048

ByteCount = 2047

Total time used:2 seconds ...

Command GetImageData Executed ...

28. On the Android device you should see surveillance photos appear along with the current surveillance frame number associated with the photo.

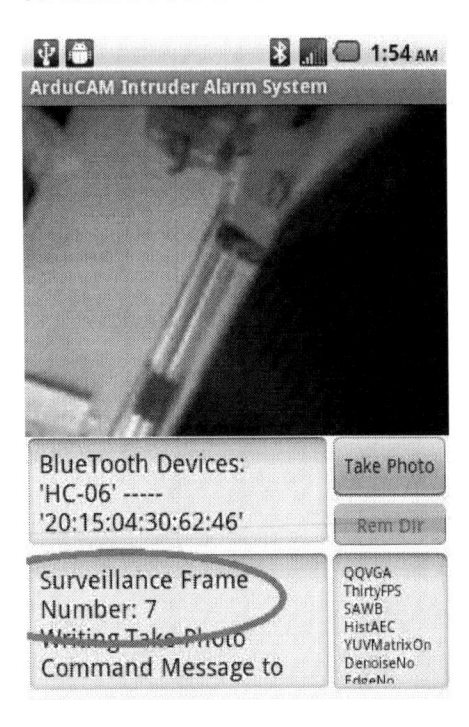

29. By default the surveillance frames are not saved. However you can save the surveillance frames to the Android device's file system. Turn off the surveillance. Next, go to the main menu

and select "Android Storage Settings". Select the "Save Images to Android" selection to start saving the incoming photo images.

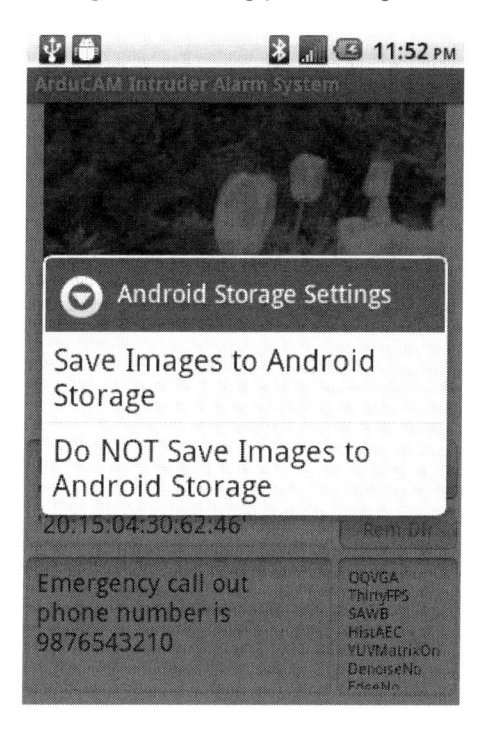

30. Turn on the surveillance. Now, each photo that is displayed will be saved to the Android's local storage in the "ArduinoCameraPics" album in the Pictures directory. The name of the file the image is saved under is also shown in the Debug Window.

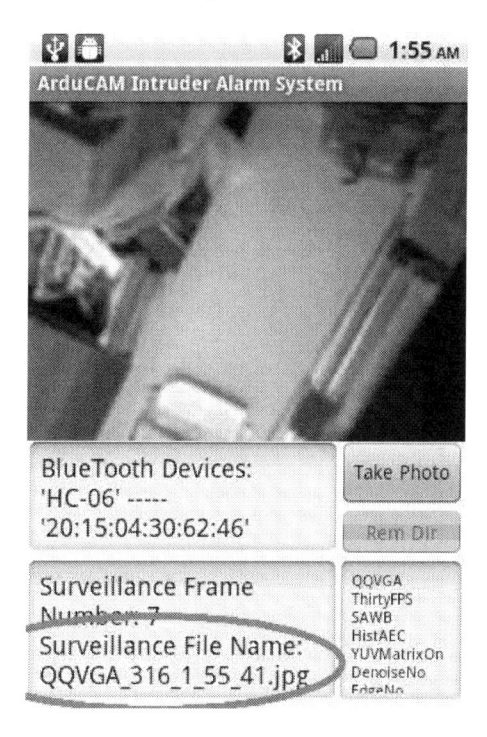

31. Turn off the surveillance.

246

32. You can load in the surveillance images that were just saved by going to the main menu and selecting the "Load JPEG Image From Android" item.

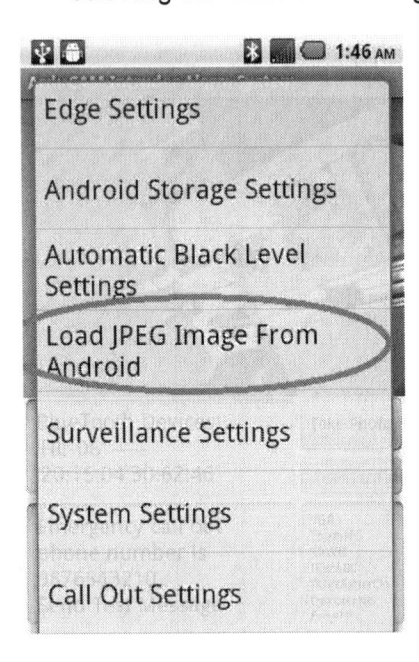

33. This brings up a list of JPEG files located in the ArduinoCameraPics directory that is used to store the incoming images from the camera. Click on a selection to load in the image for display on the Android device.

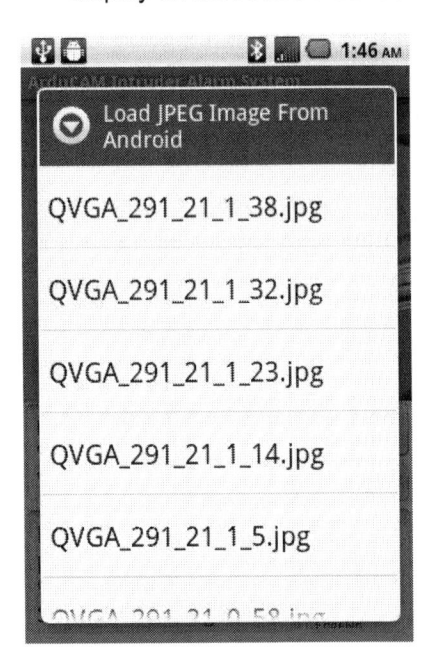

34. Now let's start up the intruder alarm. Go to the main menu and select the "System Settings" selection. Another menu should come up. Select "Arm System" to arm the intruder alarm system.

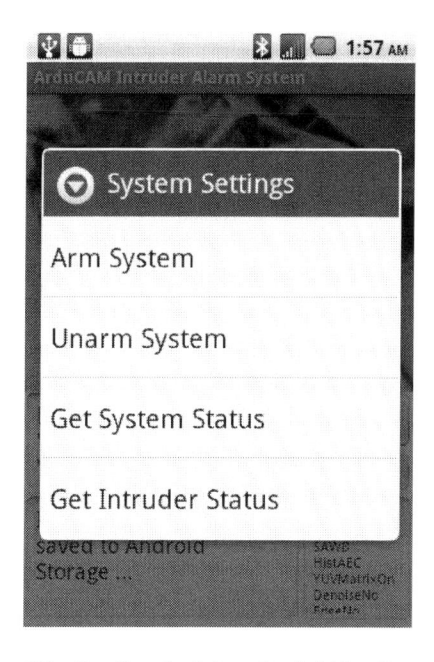

35. On the Arduino Serial Monitor you should see the "SystemStart" command being received and should see the confirmation that the intruder alarm system has been turned on.

Raw Command from Android: SystemStart

Command: SystemStart

System is Activated ....

Command SystemStart Executed ...

36. Make sure you have the motion sensor set up so that it only detects motion you want it to. For example, I used the plastic casing of a box of dental floss that was cut in half to hold the motion sensor.

37. Now, its time to test the intruder alarm. Pass your hand in front of the motion sensor. You should see a message from the Arduino. The following message indicates that the value of the sensor is 1 and the total number of intrusions detected since this Arduino program has started running is 1.

IR SENSOR(Value,NumberHits)= 1 , 1

38. On the Android side by default the call out to the emergency number is triggered by the motion alert message from the Arduino. The Android will attempt to call out to the number specified in the Arduino code as the emergency contact number.

39. Next, click on "End" to end the call.

40. What you should notice when you get back to the main Android screen is that video surveillance has automatically started. If you have selected to save images to the Android file system then the surveillance filename will also be shown in the Debug Window.

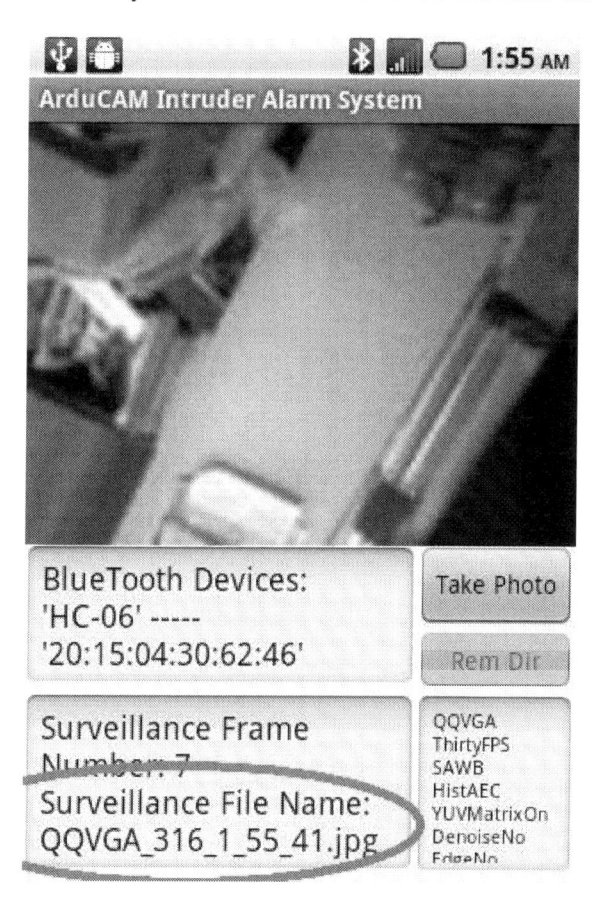

41. Disarm the system by going back to the "System Arming" menu and selecting the "Disarm System" menu selection. You should also see the Arduino side receiving this "SystemStop" message and disarming the intruder alarm system.

Raw Command from Android: SystemStop

Command is SystemStop ....

Command: SystemStop

System is DeActivated and IR Sensor is Reset ....

Command SystemStop has been executed ...

41. Now, go back to the menu and select the "Emergency Call Out" submenu. Turn off the emergency call out feature by selecting the "Deactivate Emergency Call Out" option.

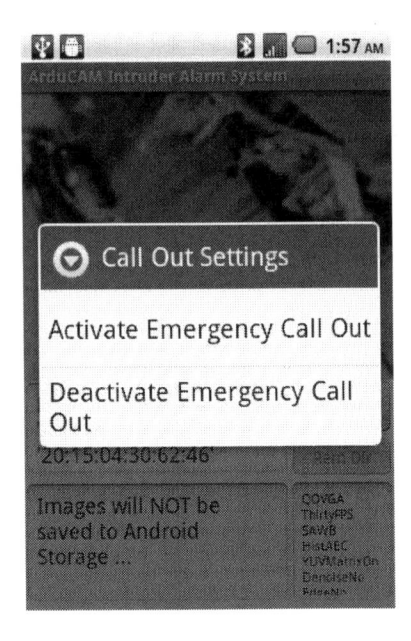

42. Arm the system again and wait for the response from the Arduino acknowledging the command.

43. Trip the motion sensor and this time you will get a vocalized audio warning of an intruder alert as well as a sound effect that continuously plays until you touch the Output Message Window or the Debug Message Window. Video surveillance will also start automatically.

44. The Serial Monitor should also show an updated count of the total number of intruders detected while the Arduino has been powered up or after the last system reset. Number of intruders should be 2.

IR SENSOR(Value,NumberHits)= 1 , 2

45. Finally, go to the main menu and select the "System Settings" menu. Select "Unarm System" to disarm the intruder alarm system. You should see the following on the Arduino's Serial Monitor. The intruder alarm system is now turned off.

Raw Command from Android: SystemStop

Command: SystemStop

System is DeActivated and IR Sensor is Reset ....

Command SystemStop Executed ...

Chapter 8

Deploying your Wireless Intruder Alarm and Surveillance System

This chapter will discuss deployment of your new intruder alarm/surveillance system. I first discuss the easily deployment method which I call the portable cardboard box installation method. Next, I discuss power supply options followed by other considerations such as how to make a more permanent and durable alarm system

Portable Cardboard Box Installation

One inexpensive way to deploy your new intruder alarm system is to just put everything in a card board box with the motion detector and camera outside of it pointing in the same direction. See Figure 8-1.

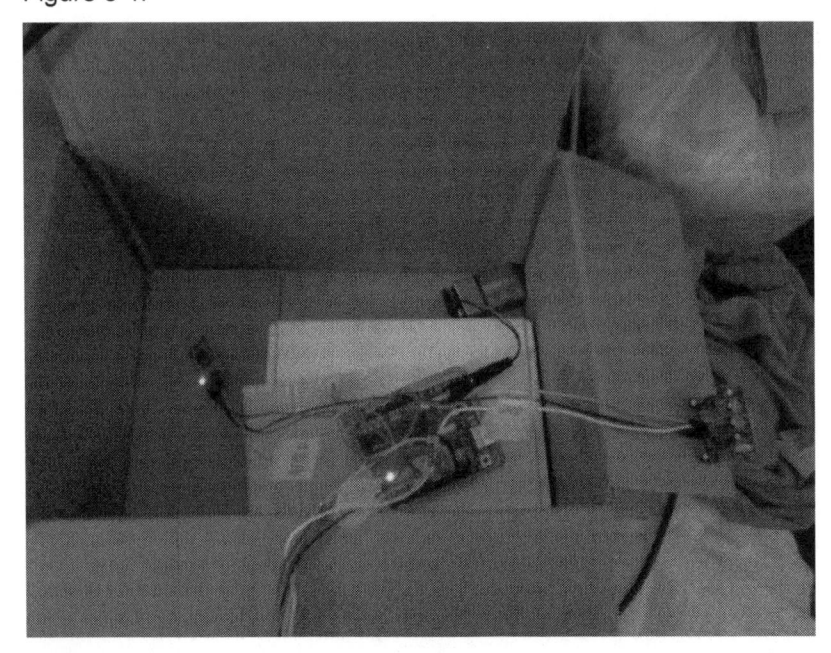

Figure 8-1. Simple portable card board box deployment method.

The camera is pointed in the same direction as the motion sensor so that when motion is detected and the camera starts taking photos the image of the intruder will be captured. See Figure 8-2.

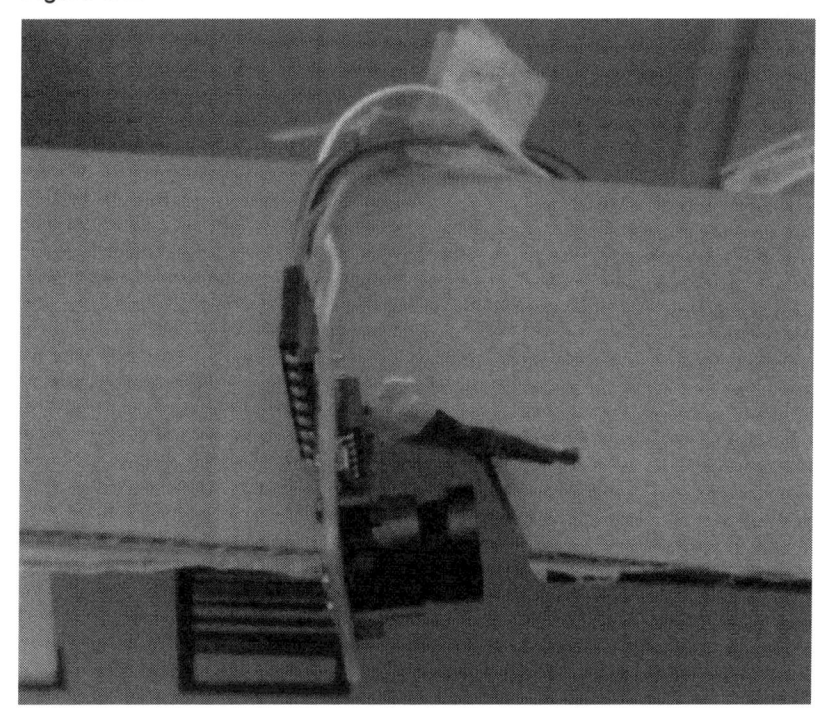

Figure 8-2. Close up of the ArduCAM Mini camera deployed on a card board box

The main idea is that you have most of the alarm system inside the box including the Arduino and just have those components outside of the box that require exposure to the outside environment such as the camera and the motion sensor. You can hide the box behind other objects such as plants, books, etc. and just have the camera and motion sensor portions of the system visible in order to take photos and to detect intruders.

Power Supply

This section covers the various methods to power your new intruder alarm system. Besides being powered by the USB port on a computer there are two other ways to power your alarm by using a 9-volt battery or an AC outlet power adapter.

9-Volt Battery

A 9-volt battery can be used to power an Arduino. However, I recommend an AC adapter especially if you are using a camera. If you are getting malfunctions or strange behavior when using a 9-volt battery to power the alarm system try using an AC adapter instead. See Figure 8-3.

254

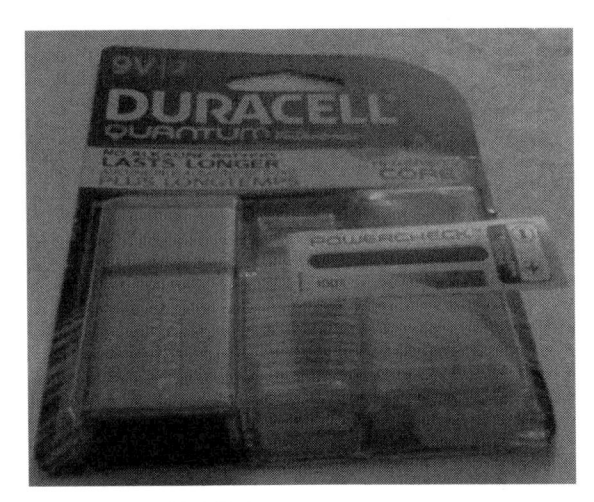

Figure 8-3. 9 Volt Battery

9 Volt DC to AC Power Adapter

An AC power adapter will provide your Arduino with continuous power from your home's power outlet. Just stick the DC plug into the Arduino and the AC prongs into your home's power outlet. See Figure 8-4.

Figure 8-4. 9 volt DC to AC power adapter

Arduino Case

A plastic case that encloses the Arduino will help keep the dust out while keeping access to all the pins and other plugs on the Arduino. There are several different companies that make an Arduino UNO specific case and one of these cases is shown in Figure 8-5.

Figure 8-5. Arduino Case

Custom Shield Overview

The Arduino prototype shield fits on top of the Arduino and has areas on it where you can solder components and then connect them to the pins going to the Arduino.

If you want to make an intruder alarm system that is more permanent then you can use an Arduino prototype shield to solder permanently wires to the prototype shield. There are also pins on the bottom of the shield that fits on top of the Arduino Uno board. See Figure 8-6.

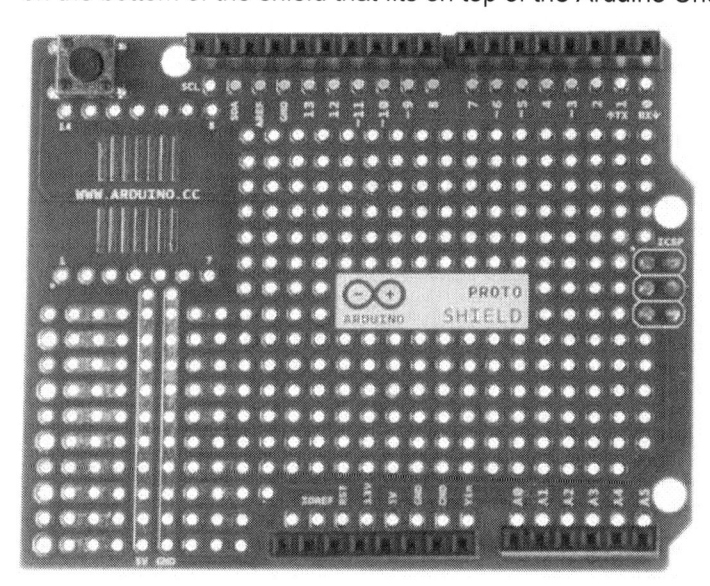

Figure 8-6. Arduino Prototype Shield

Tools for Creating a Security/Surveillance Shield

In order to attach wires to the shield you will need a soldering iron and solder. A soldering iron heats up the shield, the wire and the solder so that the solder melts and the wire is attached to the shield. A typical soldering iron is shown in Figure 8-7.

Figure 8-7. Soldering Iron

The solder itself is generally silver in color and sold in rolled coils in tubes like that shown in Figure 8-8.

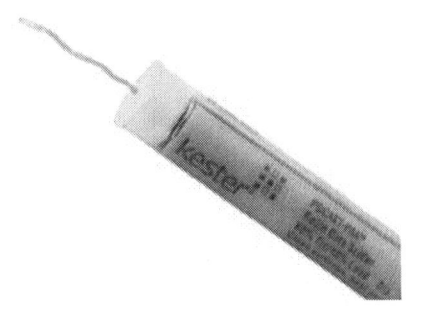

Figure 8-8. Solder

In addition to a soldering iron you will need something to hold the iron when it is hot such as a specially designed device such as shown in Figure 8-9.

Figure 8-9. Holder for soldering iron

There are also other prototype circuit boards that you can use such as those shown in Figure 8-10. These boards would also serve as good practice platforms for those not familiar with soldering electrical components.

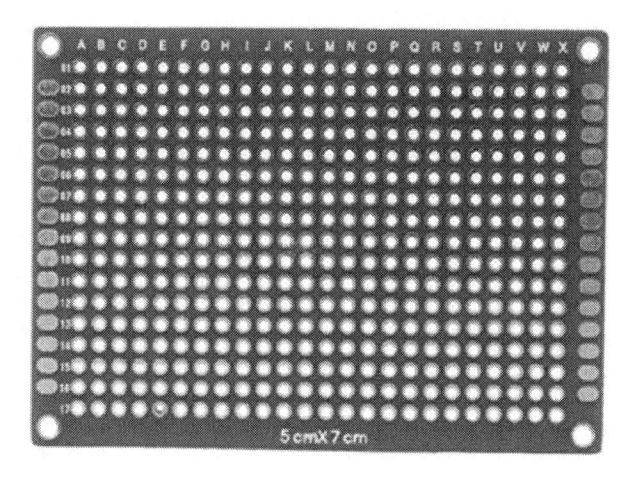

Figure 8-10. Single prototype circuit board

Printed in Great Britain
by Amazon